ENVIRONMENTAL ISSUES IN INDIA

ENVIRONMENTAL ISSUES IN INDIA

Edited by
Dr. Bhavana Trivedi
Senior Lecturer
Dept. of Political Science
Arya Mahila Degree College
Varanasi
&
Dr. Sangeeta Jain
Lecturer
Dept. of English
D.A.V. Degree College
Varanasi

DPH
DISCOVERY
PUBLISHING HOUSE

First Published – 2007

Reprinted – 2025

ISBN: 978-81-8356-224-9

Environmental Issues in India

Published by:

DISCOVERY PUBLISHING HOUSE
4383/4B, Ansari Road, Darya Ganj
New Delhi-110 002 (India)
Phone: +91-11-23279245; 23253475; 43596065
Mobile: +91 9811179893 / +91 9871656464
E-mail: discoverybooksindia@gmail.com
orderdphbooks@gmail.com
namitwasan9@gmail.com
web: www.discoverypublishinggroup.com

Printed at:
Infinity Imaging Systems
Delhi

Dedication

A Step in Ganga Seva Nidhi's Journey to Environmental Renaissance

काशी हिन्दू विश्वविद्यालय
प्रौद्योगिकी संस्थान
वाराणसी-२२१ ००५
BANARAS HINDU UNIVERSITY
INSTITUTE OF TECHNOLOGY
VARANASI–221 005

BLESSINGS

The earth's environment has been continuously changing through natural and anthropogenic activities. Increasing use of natural resources (fuels, minerals, land, water, etc.) by the growing human population during the post-industrial revolution has greatly speeded up this change. Population growth has been the single most important factor which has led to depletion of natural resources and dramatic increase in the output of a variety of wastes. It is high time to recognize that the human over population and its spiraling needs, wants and greed are the most serious problems. Most conflicts between humanity and earth's environment are based towards material well-being. People of developing countries are worried about their survival and those in developed countries about the comfort and accumulation of wealth. Thus at the global level there is a perceptible clash of need and greed. It is high time for one and all to realize that the pursuits of economic development and environmental conservation can proceed hand in hand as equals. The sooner we start giving top priority to the welfare of the ecosystems, the better it will be for all of us, because without healthy ecosystems, and economic gains that accrue can only be temporary and unsustainable. The primary objective of sustainable development should be to achieve healthy functioning of the ecophore and the economic consideration should be secondary.

The health and well-being of the planet cannot be safeguard by science and technology alone. For this dialogue and interaction between scientists and technologists from different disciplines, and with economists, sociologists , demographers, law experts, political scientists, philosophers, etc. are essential. Scientists and technologists must express their concerns to social scientists to while the latter should make environmentalists understand their language.

From the turn of the 20th Century, the emphasis on environmental issues has changed markedly from end-of-the-pipe treatment approach to reduction and resource, to recycle and reuse, to global environmental problems, to life-style changes, to need to involve one and all. This compilation of articles by various scholars is an attempt to wed the basics concepts of Indian philosophy , ethics, law, education, constitutional provisions, societal needs, politics, bureaucratic mentality, literature management with elementary concepts of environmental science and ecology. All chapters are written by persons not directly working in the area of pollution control, hence, a new direction is expected from the authors.

This book, will be useful for all those interested in tackling environmental issues to some extent through non-technical and non-scientific approach. I congratulate all the authors for their contributions and also the editors for coming out with the idea of compiling the views of distinguished authors in a single volume.

Professor S.N. Upadhyay
FNASc, FIE
Director

Foreword

The problem of environment is a big question today when it has reached the threatening stage of human survival. It is not a making of one day or in the recent past. In fact, it is the result of the very basic postulates on the basis of which the modern society exists-in essence the result on which the whole project of modernity rests. The central idea in the whole concept of modernity is the development of scientific spirit leading to the domination/mastery of man over nature to realize the full potential of nature for the development, happiness and creation of a just, human society. It establishes man not just as one among the many species of this diverse beautiful planet which has organic biological relationship in terms of its evolution but as one endowed with the physical, intellectual, knowledge driven, scientifically superior position and power to exploit other species for the benefit (?), welfare and advancement of human beings. As things progressed as a result of this scientific, mechanistic, technocratic, view points which shaped the powers of developed, industrialized west the gradual exploitation of the men and resources of the developing world for the creation of a just, happy consumerist social order of the west through the reckless exploitation of nature is now manifesting itself in the ecological imbalances in the form of Katrina, T-tsunami, floods shortage of drinking water etc. Earlier the price of development was paid by the developing world for the benefit of the developed world in terms of blind exploitation of its natural resources. When the same model of western development was adopted by other newly independent countries of the rest of the world, the problem of environment became more apparent, its impact more immediate and long lasting. It brought in miseries for a large chunk of human beings through deforestation, eviction of the local population for the construction of industrial units, dams for the so called benefit of the large mass of people .How much it benefited remains a big question but in its trail left behind the question of landless laborers, settlement of ousters, the very question of their existence and relevance for the larger society which believed in aping the living standard of the industrialized west.

The technological advances have brought in the concept and realty of 'one world' that the problems emanating in one part is going to have impact on others as well. The wall of isolation has not only thinned but collapsed as far as environment and environmental problems are concerned. Once the environmental and ecological disasters, whether man made or natural started affecting others the need for protecting environment, enduring systems of sustainable development not only gained currency but occupied the minds of the right thinking people throughout the globe.

Whatever the situation, it is a commonplace knowledge that environment has to be protected if the human race has to survive. Already so many species have become extinct since the creation of the earth. There is no guarantee that the human species will exist for ever. However, one thing remarkable about human beings is that they have the capacity and intellect to evolve an order in which they can not only save themselves but other species which are endangered and going to become extinct. How this can be done is a major question which must be thought of and attempted both globally and locally by all right thinking people. Does it mean reversing the till now adopted scientific and industrial action and bans of development? This in turn means questioning not only the basis of acquiring knowledge and its use but also redefining man-nature relationship. This is a tall order. This requires harvesting the intellectual capacity, philosophical wisdom, respecting the age old traditions of man-nature relationship in societies, global concerted action for the agenda making for governments, MNCs, governing elites, creation of pressures by the civil society locally, nationally and internationally. There in an urgent need to integrate the local with the global in managing environment. The pronouncements, plans and concrete action of governments and governing elites for keeping the environment safe need careful examination and scrutiny. There is a need for concerted autonomous action on the part of scientists and social scientists to make independent findings and make the citizens informed about it. This will bring back the role of the social services in policy making, its monitoring and informing the people of the outcome. This again is a difficult task but it is a task which can be ignored further at our own peril–the survival of the human species itself.

The present volume is a welcome addition to the already existing plethora of literature on environment. Since the effort of the scientists, technologists, philosophers, have gone into the production of this volume, it will add not only to our understanding of the problem but also help us in finding some solution for the local problems. In this sense, the effort of the contributors in this volume which encompasses the global vision with local problems is both welcome and laudable.

Prof. Rajani Ranjan Jha
Ex. Head
Department of Political Science
Banaras Hindu University
Varanasi

Acknowledgment

Varanasi referred as Benaras or Kashi i.e. 'Luminous', 'radiant', 'illuminated' is a blessed city–a city redeeming all human sins and dross, liberating the human soul. The celestial abode of Lord Shiva, the decorated crescent-form vessel of the nectar of Ganga, Varanasi is visualized as a city of great antiquity since time immemorial.

The Ganga, the divine fountain that is flowing from Goumukh-the holy cave in the Himalayas to Ganga Sagar-the Bay of Bengal, is the heart throb of India. The Ganga has been the sustainer and cherisher of Indo-Gangetic civilization on its banks. The colossal flights of stone steps, great stone piers of the majestic Palaces and Mansions, multitudes of devotees are extant and living testimony to the eternal glory and grandeur of this spiritual city.

On this sacred land of Kashi, the Divine blessing and inspiration of Brahma Swaroop Sri Sadguru Bapuji Maharaj, sowed the seeds of an organization viz Ganga Seva Nidhi(GSN) to rejuvenate the declining glory of the timeless city Kashi and to preserve the purity, sanctity and divinity of Ganga and its most picturesque world famous Ghats. GSN's mission is to bring back the glorious splendour of the past through its multifarious activities and efforts and to make people aware of the priceless cultural and spiritual legacy of India.

This book *Environmental Issues* is the result of GSN's deep rooted concern for deteriorating environmental phenomena of the world. Leading academicians and experts from various disciplines have contributed their expertise in the form of articles to make this venture estimable. We are highly indebted and thankful to the intelligentsia of authors as contributors. Paucity of words binds our deep and sincere emotions for the inspiration of Mr. Narayan Singh Bhayal, Mr. Satyendra Mishra, Mr. Shiv Kumar Mishra , Mr.Amitabh Bhattacharya , Mr.Indu Shekhar, and Mr. Bhola Nath Mishra.

My sincere thanks goes to my husband Mr. Rajeev Trivedi for his kindness and patience during this strenuous task.

I express my heartiest thanks to my co-editor Dr. Sangeeta Jain for her indispensable co-operation in completion of this work.

I am also thankful to Mr. Tilak Vasan, proprietor, Discovery Publications, New Delhi for his co-operative approach and positive response in publication of this book.

Last but not the least, I also express my deep sense of gratitude to all the members and well wishers of Ganga Seva Nidhi, Varanasi.

(Dr. Bhavana Trivedi)

Secretary

Ganga Seva Nidhi

Contents

Contributors

Dr. A.N. Tripathi, Professor, Electrical Engineering, Institute of Technology, Banaras Hindu University. Varanasi.

Dr. Anup Kumar Mishra, Lecturer in Economics, D.A.V. Degree College, Varanasi.

Dr. Aparna Joshi, Lecturer, Department of Sociology, G.D.H.G.P.G. College, Moradabad.

Arun Kumar Upadhyaya, Reader, Political Science, Government P. G. College, Chandauli.

Mr. Arvind Srivastava, Advocate, High Court, Allahabad.

Dr. Bhavana Trivedi, Senior Lecturer in Political Science, Arya Mahila Degree College, Varanasi.

Dr. C.M. Jariwala, Retd. Ex-Dean & Head , Faculty of Law, Banaras Hindu University. Varanasi.

Dr. Chandrakala Padia, Professor, Department of Political Science, Banaras Hindu University and Co-ordinator Centre For Women's Studies & Development, Banaras Hindu University, Varanasi.

Dr. Devendra Mohan, Reader, Environmental Engineering Division, Department of Civil Engineering, Institute of Technology, Banaras Hindu Unitersity, Varanasi.

Dharmendra Kumar Mishra, Senior Lecturer, Department of Law, M.B.S.P.G. College, Gangapur, Varanasi.

Prof. G. S. Kushwaha, Retd. Head & Dean, Faculty of Social Science, Banaras Hindu Unitersity, Varanasi.

Dr. Girish Chandra Pande, Senior Lecturer, Department of Sociology, K.G.K.P.G. College, Moradabad.

Dr. Kalpana Chakravorty, Ph.D. in Psychology, Senior Teacher, Sunbeam School, Varanasi.

Dr. Krishna Banerjee, Professor of English, Women's College, Banaras Hindu University, Varanasi.

Dr. Niharika Lal, Senior Lecturer, Department of English, Vasant Kanya Mahavidyalaya, Kamachha, Varanasi.

Dr. Priyankar Upadhyaya, Professor of Political Science, Co-ordinator, Malviya Centre For Peace Research, Banaras Hindu University, Varanasi.

Dr. Pushpa Tiwari, Reader, Department of Ancient History, Culture and Archaeology, University of Allahabad, Allahabad.

Dr. Rachana Srivastava, Reader & Head, Department of Political Science, Vasant Kanay Mahavidyalaya, Kamachha, Varanasi.

Rajendra Bharti, Senior Lecturer, Department of Political Science, Simdega College, Ranchi University.

Dr. Rajeshwar Upadhyaya, Professor, Ex-Dean, Faculty of Education, Banaras Hindu Unitersity, Varanasi.

Dr. Renu Srivastava, Reader, Department of Political Science, Vasant Kanya Mahavidyalaya, Kamachha, Varanasi.

Dr. Sangeeta Jain, Lecturer in the Department of English, D.A.V. Degree College, Varanasi.

Dr. Sanjeev Kumar, Senior Lecturer, Department. of History, Vasant College, Rajghat, Varanasi.

Mr. Shardul Chaubey, Senior Lecturer, Department of Media Studies, School of Management Sciences, Varanasi.

Dr. Suman Jain, Senior Lecturer, Department of Hindi, Vasant Kanya Mahavidyalaya, Kamachha, Varanasi.

Dr. Sunil Kumar Singh, Reader, Faculty of Education, Banaras Hindu University, Varanasi.

Dr. Tarun Kumar, Senior Lecturer in Political Science, L.B.S.P.G. College, Mughalsarai, Chandauli.

Introduction

The environmental hazards and challenges confronting the post-world war period have turned out a greater menace to the survival of living organisms than at anytime earlier in human history. The subtle enormity and magnitude of insensitive needless human acts and activities, unplanned developmental accomplishments prove to be detrimental to the ecological balance of the world.

Nineties heralded the economic and political globalization giving an upsurge to the environment oriented concerns and turning them into a cross-border, cross-political, cross-national, one of the major issues of the 21st century. Pressing and captivating issues like global warming, holes in ozone layer, acid rain, toxic chemicals and biological diversity, deforestation and desertification, river and ocean degradation, air, water, soil, noise pollution need to be tackled collectively by the nations of the world.

Post world war debates on environment engender divergent and convergent ideological variations. Ecocentric and Technocentric Environmentalism, Bioregionalism, Ecofeminism and Feminist Environmentalism, Ecocriticism, Environmental Management, Green Management, Ecodemocracy, Ecoactivism, Ecolitigation , Environmental Economics Ecolabeling etc. are some of the latest environment oriented concepts and assumptions. Here an effort has been made to throw light upon the varied concepts related to environment. An overview of global environmental issues, their effects, the Indian environment scenario also forms a section of this introductory chapter.

I

Ecocentric mode is an imaginative attempt to place the ecosystem at the centre rather than humanity. Man's inquisitive tendency led him to explore the hidden treasures and resources of Nature and to utilize them for the benefit of humanity without considering the repercussions of endless and ignorant utilization. This irrational exploration of the clandestine natural wealth resulted in degenerative succession of distorted harmonies causing self-destruction by nature-destruction. Thus the ecocentric environmental philosophy demands a permanent and stable behavioral code of conduct for reverence, responsibility and care of nature and natural resources. Various studies (e.g.Burch 1971) on Ecocentric philosophy reveal that it is an oblique offshoot of and nourished by the American romantic transcendentalist philosophies of mid-nineteenth century that advocate for a powerful cordial relation and association between nature and society. For building a strong cultivated protective view towards ecological system it is necessary for human beings to realize the potential of nature and ecosystem as driving force conducive to the development of a sublimated human soul. Walt Whitman, American poet and philosopher necessitated the need to understand human responsibility towards nature and mankind. The transcendentalist philosophers state that city is the domicile of competitions, harshness, rigidity therefore it contrasts nature which is uncompetitive, cool, pure, aesthetic, mild. Similar contrastive elements of nature and city ambience are reflected in oriental philosophy as Rabindra Nath Tagore observed that contemporary western civilization is built of brick, iron and wood and it is rooted in the city. But Indian civilization is born of nature as it locates the source of regeneration both spiritual and material in the forest.

Ecocentric mode developed into two subsequent lines of thought viz "Bioethic" and "Self-reliant". Bioethical notion propounded by 'nature-moralists' demands the safeguarding of nature and eco system as nature has right to itself. Nature and its wilderness are integral companions to man because thousands of overtired, exhausted, fed up people want to return to nature to soothe their fatigued soul. Muir and other nature moralists have observed that several people are beginning to find out that returning to mountains is akin to going back home. They feel wilderness is a necessity and that mountains, parks, fields greenery are useful as 'fountains of life' as well as source of timber and water for irrigation.

The International Union for the Conservation of Nature and World Wild Fund propagated the bioethical view to preserve forest areas, national parks and natural environment, plants and animals and the establishment of world Heritage Trust is also an initiative in this direction. Muir (1945), Leopold (1949), P. Anderson (1971), Brooks, J. Olson, Sauer(1971), Dasmann, Ethrenfeld (1972), Shepard and Shepard (1973) and Goetham (1974) in the United States; Brinkhurst and Chant (1971) Littlejohn and Pimlott (1971) and Livingston(1972) in Canada; and Nicholson(1970,1973),Fraser-Darling(1971),and Allsop(1972) in Britain are notable nature moralists advocating for Bioethical view. J. Stone (1972) an environmental lawyer supported a common law of bioethic rights, a concept first suggested by Supreme Court Justice Douglas who stated "Contemporary public concern for protecting nature's ecological equilibrium should lead to a conferral of standing upon environmental objects to see for their own preservation."

The Self Reliant Community is an extension of Ecocentric mode which puts the nature and Ecosystem at centre. It advocates for the establishment of small self sustaining communities where natural resources are utilized judiciously and thoughtfully for social objectives. Intellectual theorists like Peter Kropotkin, Ebenezer Howard, Patrick Geddes, Patrick Abercrombie, Lewis Mumford, Paul Goodman, Theodore Roszak, Murray Bookchin developed the idea of small self dependent communities.

Rapid industrialization and urbanization disillusioned the intelligentsia of high profile facets and brought in the forefront the dehumanizing and dissocializing effects, the artificial and alienating ambience of metropolis, the intellectually deadening impression of specialized occupational courses and the wastage of scarce human capabilities and energies. The side effects and counter effects led the social thinkers to envisage about the measurement that compares the economic effects with social consequences. Peter Kropotkin (Gould, 1974; Ward, 1974, Woodcock, 1974) assessed the patterns of industrial development and its social repercussions. He developed his view at the juncture when the political debates centered upon the effects of Industrial revolution were visible upon society triggering the report of the poor law commission and the writings of Karl Marx. Kropotkin favoured the view of a "simple life based upon limited expectations where there was little demand and need for constant material progress" . He assured that happiness could be attained through a "regional pattern of decentralized communities". These small community groups would have agriculture and small industrial enterprises where the people would have necessary education and occupation demanding manual labour with creative intellectual activity. This community system would have segregated work culture for employers and employees. Consequently this self sustained community system was thought to relieve and release man from soul destroying burdens of monotonous, exhaustive labour. The rations and concepts of "Economics" of permanence developed by Schumacher (1973) the Convivial Community of Pirages and Ehrlich (1979) the 'Visionary Commonwealth' of Roszak (1973) and 'Monastic Commune' of Heilbroner (1974) try to establish a new empathetic social order beginning from neighbourhood to nation. These notions will support environmental protection as they would return to nature, live amidst natural surroundings and understand the significant contribution of ecosystem in developing a pro-eco mindset, But the only drawback is that the self Reliant community theme is apolitical as it removes ecocentric community from practical political arena without which it will have no power to change and sustain the existing social order.

II

Technocentric mode is the reverse stand of Ecocentric mode which shows the sense of wonder, reverence and moral obligation towards Ecosphere Technocentric ideology downplays the ecocentric convictions and reveres Man's abilities and capabilities to control phenomena. In other words "The technocentric mode is identified by rationality, 'the objective' appraisal of means to achieve given goals, by managerial efficiency, the application of organizational and productive techniques that produce the most for the best effort, and by a sense of optimism and faith in the ability of man to understand and control physical, biological and social processes for the benefit of present and future generations." 'Rationality', "Managerial efficiency", "Control" are the keywords of Technocentric stream that is directed towards progress. By and large the technocentric viewpoint holds the view that greatest possible use of the natural resources with growing specialization of resource management techniques would lead to the ultimate progress and success. Analysts (Hays, 1959; Weisberg, 1971; Leiss, 1972) say that technocentrism is a productive action of a particular kind of economic and political existence emerging through the amalgamation of economic and political power during the first American conservation movement at the turn of this century.

III

The ideology of Bioregionalism states that the environmental crisis starts at home and depends on individualistic perception of the concerned space in the world. "All the planet wide pollution problems originate in some bioregion," Says Berg. That is each region in the world has its own relative environmental problem, The responsibility resides on the shoulders of the natives to eliminate the cause of the hazards. If inhabitants strike a five-finger chord by promoting more community services like setting up community gardens in the nearby areas, establishing more renewable energy sources, arranging easily accessible public transportation and improvising methods and means for sustainable agriculture, certain problems might be controlled up to an extant e.g. rain forest will stop getting chopped down in direct proportion. Bioregionalists who make these suggestions, live within the existing communities and regions where they face certain regional environmental problems.

Gaia (Greek goddess of Earth) hypothesis, first proposed in 1972 established a reciprocal relation between evolution of species and evolution of their physical environment. British Chemist James Lovelock who worked with a NASA team investigating the possibility of life on mars, held that "the evolution of the species of living organisms is so closely coupled with the evolution of their physical and chemical environment that together they constitute a single and indivisible evolutionary process." The idea inferred from Lovelock's observation is that the external atmosphere, despite being composed of an unstable mixture of reactive gases such as oxygen and methane, for the entire 5,500 million years since the inception of life has remained remarkably stable. He hypothesized that "living matter, the air, the oceans, the land surface, were parts of a giant system which seemed to exhibit the behaviour of a living creature." Eventually these hypothetical formulations lead to figure out that all living creatures and the natural elements of earth are in a symbiotic relationship and exhibit reciprocal complementary characteristics. The sustenance of Goddess of Earth i.e. Gaia will sustain its creatures. Whether metaphor or reality, Gaia is in serious trouble because the dynamic balance of the biosphere is being radically disrupted by human activities such as the destruction of tropical forests, fossils, fuel consumption, and the release of chlorofluoro or carbons. The biosphere is under greater stress than at any other time in human history. Our choice is between living and surviving.

IV

Ecofeminism and Feminist Environmentalism seek new dimensions of both feminism and environmentalism. Ecofeminist stand conceptualizes the link between women and the environment primarily in ideological terms where as feminist environmentalism finds this link rooted in the material reality and requirements of men and women. As a body of thought ecofeminism focuses on various entangled threads in the relationship of gender and nature. It postulates that the domination and exploitation of nature and women have deep interconnections with culture and men. Patriarchal thought assigns nature and women an inferior status to culture and men respectively, therefore both are exploited. Feminist movement and environmental movement both demand for egalitarian, non-hierarchical systems. The commonalities of both the movements unite them to evolve a common perspective for both. Feminist environmentalism makes an explicit analysis of "material reality", the structural causes of both environmental degradation and women exploitation. Society is structured on the basis of gender and class, division of labour and distribution of property and power. These factors not only divide the mindset but prepare their responses to environmental as well as women problems. e.g. poor peasant and tribal women fetch fuel, fodder and water and have been the main cultivators in hills. On one hand they are exploited on homefront and they exploit natural phenomena but on the other hand they have experiential learning of species, varieties and the process of natural regeneration. Thus they are vehicles of both degradation and regeneration and are "repositories of knowledge." It alternatively prepares the feminist perspective about environment, structured on the basis of specific class and gender, division and distribution of labour. This ideological construction about the interdependency of gender and nature generates the philosophy of feminist environmentalism.

V

In modern Indian intellectual tradition Gandhian Philosophy also contributed in the area to environmentalism as Gandhi said, "Man has no power to create life, therefore, he has no right to destroy life." And further he commented that "The earth has enough resources for our need, but not for our greed."

Mahatma Gandhi, a great political leader, revolutionary, great thinker and saintly figure, humanist and pacifist of international repute, was an environmentalist too. He was not convinced with western techno scientific culture and socio economic and political theories. Making a comparative analysis of west and east he stated that 'Economic imperialism of a single small island (England) keeps the whole world in chains, if a whole nation (India) of 30 crore populations will proceed on the path of such economic exploitation it will eat up the whole world like locusts. Not only did he exhort the people to see the adverse dink between the western civilization and the threat to natural resources leading to what is known today as "ecological crisis" and what he called the exhaustion of earth in resources if all the countries went the way the west was going.

Gandhi's principle of nonviolence reflects his concerns for environment and its constructing constituents. He explained the inseparable relation between self realization, non-violence and biospherical egalitarianism. "His anthropocentric vision is revealed when he said, "If our sense of right and wrong had not become blunt, we would recognise that animals had rights, no less than men"; and "we should feel a more living bond between ourselves and the rest of the animate world."

Gandhi proposed a new industrial activity non-pernicious to nature. We stated "I have not contemplated, much less advised the abandonment of a single, healthy, life-giving industrial activity for the sake of hard spinning. The entire foundation of the spinning-wheel vests on the fact that there are cores of semi unemployed people in India. ... The spinning wheel is destructive of no enterprise whatever. It is life-giving activity." Thus the core of Gandhian philosophy is to establish a friendly and benevolent relation with nature and its creation–Man.

Gandhi exerted a profound and indelible influence on the leading figures in the fields of Deep ecology, Peace research and Buddhist economics apart from several other fields whereupon his philosophical thoughts left an impact. Arne Naess, Johan Galtung and E. F. Schumacher and many environmental activists readily admit their debt to Gandhi ji. Arne Naess developed the concept of Deep Ecology wherein the influence of Gandhi is also clearly visible. Naess pointed out that "a shallow but influential ecological movement and a deep but less influential one compete for our attention". By "shallow" ecological movement he meant one that "fights pollution and resource depletion in order to preserve human health and affluence." He characterized the deep ecological movement as "operating out of a deep-seated respect and veneration for ways and forms of life, and accords them an equal right to live and blossom. Shallow ecology finds that 'Natural diversity is valuable as a resource for us'. Deep ecology says that Natural diversity has it own intrinsic value. It talks of a unification with and realization of oneness with nature, interdependency of nature and man. The theory of Holism expresses that Nature and human beings are substantial parts of each other and can't survive in the absence of one. Gandhi was a visionary who articulated the deep balanced harmony with the environment before the advent of the deep ecological strands of environmental philosophy. Mahatma Gandhi's biophysical egalitarianism is clearly visible in his perceptual statement. "Man has no power to create life, therefore, he has no right to destroy life." Further, "The earth has enough resources for our need but not for our greed."

VI

The root of the prevailing ecological crisis may be dug out in religious and spiritual environment. Major religions of the world have propounded certain moral and ethical precepts and values for humanity to establish a coherent relation with natural environment. The theme of conservation closely associated with empathetic understanding of nature is the focal point of all faiths and religions.

The Latin synonym of 'religion' is 'religare', which denotes the meaning "to bind strongly". The Greek word for 'Ecology' is 'oikos' i.e. house. Both the words connote the idea of binding, attachment, interconnectedness and interdependence. This inextricable bondage of religion and ecology has to be explored and utilized as guidelines for a safer and more sustainable tomorrow.

All faiths in the world Hinduism, Islam, Christianity, Sikhism, Jainism, Buddhism, Bahai, Judaism and Taoism have their own declarations on Nature and Conservation. A bird's eye view of the fundamentals and principles of all religions are given here with a fervent hope to bring positive change in people's outlook towards nature and conservation.

The Hindu dictum declares : 'Dharantha Dharma Ucyate' – the sustainer of all species of life and maintainer of concordance between nature and human beings is dharma and the adharma is the disturbing factor of the ecology. The faith of Hinduism resides in all encompassing sovereignty of the divine force, manifesting itself in a graded scale of evolution. The *Atharva Veda* prays:

> Earth, in which lie the sea, the river and other waters,
>
> in which food and cornfields have come to be ...
>
> Earth, in which the waters, common to all,
>
> moving on all sides, flow unfailingly, day and night ...
>
> Wakeful through a long life, we shall become
>
> bearers of tribute to thee...

The *Yajurveda* lays down that "no person should kill animal helpful to all. Rather by serving them one should attain happiness". (*Yajurveda* 13.47)

Not only the Vedas, but in the later scriptures such as the Upanishads *the Puranas, the Mahabharat, the Bhagwad Geeta, the Ramayan, the Ramcharitmanas* describe the potentials of nature and express a deep reverence to nature and admit the organic relationship between man and nature. The cosmological view of the ancient scriptural Vedas, Upanishads and Puranic traditions, literary imagination enriched by meaningful symbols and idiomatic expressions revealing inexplicable relationship of people with nature have become the axis of mode of communication.

The religious practice of worshiping nature, river trees, sky, earth as prevailed in India since time immemorial is a powerful way to express deep reverence towards the nature, the creator and sustainer of human life and thereby establishing organic relationship between men and nature. The sustenance of life depends on food, rainfall, religious sacrifice performed through rituals of Vedas-the voice of God, who created the sun, the moon, the earth and the stars and all objects in the universe jointly, not individually, create the atmosphere of everything in the universe. "Tat sristva ta devam pravisat" – this Upanishadic dictum narrates that God, the creator of the whole universe entered into every little and big objects. Therefore Hindu tradition believes that there is the existence of soul in all plants and animals. Even killing the plants and animals is a kind of penance. No life is inferior to human life and all living creatures have equal rights to enjoy the bliss of universe and to play their assigned roles.

The seed scripture of humanity *Bhagvad Geeta* disseminates the message of 'conserve ecology or perish'. In the third chapter of *Geeta,* Lord Krishna and Arjuna converse about clear and precise Life Science. The preservation of ecology must be made the highest objective of life so as to give life a secured future. The ecological cycle is described in verses 3.14-1:

> "All living bodies subsist on foodgrains, which are produced from Rains, Rains are produced from performance of Yajna (sacrifice).... Regulated activities are prescribed in the Vedas, and the Vedas are directly manifested from the supreme personality of Godhead.... My dear Arjuna, one who does not follow in human life the cycle of sacrifice thus established by the Vedas certainly lives a life full of sin. Living only for the satisfaction of the senses, such a person lives in vain."

Further Lord Krishna preaches : "I become the moon and thereby supply the juice of life to all vegetables." (15.13)

The Indian cultural tradition also in several ways highlights the connection of environment and men. Evoking the filial attachment between the mother earth and himself as her son, man recites rains to be his father, prays the earth to bestow upon humanity her blessings in the form of protection, purification and rearing:

I am the son of the earth,
the rains are my father,
let him, the lord of the rain,
fill the earth for us.
Oh earth, protect us, purify us,
Let the people milk her with amity.
Oh earth,
give us sweet words.
The snowy mountain heights
and thy forests, oh earth,
shall be kind to us
and we to them."

The inseparable relation amidst various elements of nature – water, earth, air, sky and sun receives justifiable reflection in the above mentioned lucid verse highlighting the importance of environment in the Indian cultural tradition that provides ideological underpinning and legal authenticity to the present moves and measures for environmental protection and conservation. In modern times Vandana Shiva the reputed ecologist writes "For the cultures of Asia, the forest has always been a teacher, and the message of the forest has been the message of interconnectedness and diversity, renewability and sustainability, integrity and pluralism."

Indubitably, India has the pride of having on inherent 'forest culture' i.e. an Aranya Sanskrit. Saints, seers and sages lived in woods and jungles in perfect communion with forest, a replica of nature and they drew words of wisdom from it creating ancient texts called Aranyakas (forest texts). The great Sanskrit poet Kalidas,

Mahatma Gandhi and Rabindra Nath Tagore reinterpreted the concept of nature as a teacher, a model for a humane society and a harmonious man-nature relationship.

The essence of Islamic teaching is that the entire universe is God's creation. Allah makes the water flow upon the earth, upholds the heavens, makes the rainfall and keeps the boundaries between day and night. Unity trusteeship and accountability that is tawheed, Khalifa and Akfrah, the three concepts of Islam are also the pillars of the environmental ethics of Islam, constituting the basic values taught by the *Qur'an*. It is these values which led Muhammad, the Prophet of Islam say : "The world is green and beautiful and God has appointed you his stewards over it." Environmental consciousness is born when such values are adopted and become an intrinsic part of our mental and physical makeup.

The central concept of Islam is Tawheed or the unity of God. Allah is unity, and his unity is also reflected in the unity of mankind and the unity of man and nature. His trustees are responsible for maintaining the unity of his creation, the integrity of the Earth, its flora and fauna, its wild life and natural environment. Therefore Muslims believe that Islam is a middle path and they will be answerable for how they have walked this path, how they have maintained the balance and harmony in the whole of creation around them. Countless verses in the Holy Qur'an state:

"020.006 To him belongs what is in the heavens and on earth and all between them and all beneath the soil".

"002.155 Be sure we shall test you with something of fear and hunger, some less in goods or lines or the fruits (of your soil), but give glad tidies to those who patiently persevere.

"002.156 Who say when afflicted with calamity "To Allah we belong and to him is our return."

"Christians, in particular, realize that their responsibility within creation and their duty towards nature and the creator are an essential part of their faith." This is the New Year message conveyed by his Holiness the Pope in 1990, summarizing the integral accountability of Christians to the whole creation. The Orthodox Church has brought the idea of human being's responsibility towards God's creation and nature very strongly especially in

Orthodoxy and the Ecological Crisis 1990, the document produced by the Ecumenical Patriarchate. It postulates that all human beings both individually and collectively should look at the 'natural order as a sign and sacrament of God'. Contrary to it, nature, its resources, elements and creatures are taken for granted and exploited to the point of depletion and extinction. It teaches that humanity should maintain a cordial relation between God and the world as it was before in Eden. Repentance is the only step, which can unite the two landscapes – the one human, the other natural to make both the 'objects of a caring and creative effort.' The feeling of true repentance must have a clear outline of initiatory steps to reflect the ethos of Orthodox Christianity.

The World Council of Churches comprising Protestant and Orthodox Churches in its collaborative congregation of their member Churches in 1990 discussed the issues of Justice, Peace and Integrity to creation:

> "Affirmation VIII
>
> We affirm the creation as beloved of God.
>
> We affirm that the world, as God's handiwork, has its own inherent integrity; that land, water, air, forests, mountains and all creatures, including humanity, are 'good' in God's sight. The integrity of creation has a social aspect which we recognize as peace with justice, and an ecological aspect which we recognize in the self-renewing, sustainable character of natural ecosystems.
>
> We will resist the claim that anything in creation is merely a resource for human exploitation. We will resist species extinction for human benefit....
>
> Therefore we commit ourselves to be members of both the living community of creation in which we are but one species, and members of the covenant community of Christ; to be full coworkers with God, with moral responsibility ...."

One of the youngest living faiths, founded by Lord Guru Nanak, Sikhism's last Guru Gobind Singh directed the followers to revere the *Guru Granth Sahib* as the word teacher of Sikhas. The principal tenets in the religious scripture teach to see the glory of God in nature and environment Lord Guru Nanaka explained:

Nature we see

...

Nature in the whole creation

...

Nature in food, in water

...

Nature in life forms

...

All nature is yours, o powerful creator,

you command it, observe it and pervade within it.

(*Adi Granth*)

Like other religions and faiths, Sikhism also preaches the message of the "Oneness of Creation" and Almighty God created each and every particle of this universe. The five elements of nature air, water, earth, fire and space create the earth "the place for Righteous Activities." (*Guru Granth Sahib*) and a balance in all the elements in nature is mandatory for the survival of the Universe. The Gurus of Sikha religion have emphasized human beings responsibility towards the earth i.e. "you are the master of this earth and responsible for all life forms." (*Guru Granth Sahib*)

Jainism and Buddhism both the faiths have deep ecological concerns in their teachings. Jainism, consolidated by the twenty-fourth Tirtharkara Mahavira, believes that the attainment of the supreme stage of personal development is possible with adherence to the three jewels (ratna-traya) namely enlightened worldview (Samyak Darshan), true knowledge (Samyak Gyan) right conduct (Samyak Charitra). Non-violence, Truthfulness, Non-stealing, Chastity, Non-materialism are the five small vows. (Anuvratas)

Ecological harmony and nonviolence form a vital part of Jain philosophy and its spiritual quest. The central message and motif of the Jain Perspective depend on its respect for all living and non living things, its commitment to the progress of human civilization along with the preservation of natural environment. The principle of ahimsa – "Ahimsa parmodharmah" (Non-violence is the supreme religion) is not only towards human beings but towards all nature. The scriptures explain. "All the Ashats (Venerable Ones) of the past, present and future discourse, counsel, proclaim, propose and prescribe thus in unison : do not injure, abuse, oppress, enslave, insult, torment, torture or kill any creature or living being"

The principle of "Parasparopagraho jivanam" (interdependence) recognizes the interdependence of natural phenomena which forms the fundamental principle of Ecology. This scriptural aphorism states that all life is bound together by natural support and interdependence. All the facets and dimensions of Existence belong to one another and have a physical as well as a metaphysical relationship. Life of all living and non living beings is a continuous struggle of coexistence and cooperation in a 'universe teeming with interdependent constituents.'

Buddhism may also be recognized as an ecological religion or a Religious Ecology. Lord Buddha unfolded the mystery of existence that no one can exist in isolation and 'an isolated entity is an illusion.' There is an inseparable link in the seemingly autonomous elements of the Whole.

Several Buddhist monks such as His Holiness the Dalai Lama, Venerable Thick Nhat Hanh, Venerable Kim Teng and Venerable Phra Phrachah emphasize the natural relationship between deep ecology and Buddism. Vietnamese monk Venerable Thick Nhat Hanh has said. "Buddhists believe that the reality of the inter connectedness of human beings, society and Nature will reveal itself more and more to us as we gradually recover ... efforts to change the environment and to change oneself are both necessary."

Buddhists ideals relocate natural word within contexts-Nature is our oracle that teaches us to accept the multitudinous changes which occur during the transition from birth to death. Buddha believed that Nature is the permanent provider of Wisdom and looked at nature as a spiritual force for meditation practice. Shantineda in eighth century India, states dwelling in nature is more suitable to spirituality than living in a monastery or town : "When shall I come to dwell in forests Amongst the deer, the birds and the trees, that say nothing unpleasant and are delightful to associate with" (A Guide to the Bodhisattva's Way of Life). Lord

Buddha taught that the natural functions of the forest are conductive to ecological balance, natural harmony and survival of environment.

Bahai faith declares that, "... even as the human body in this world, which is outwardly composed of different links and organs, is in reality a closely integrated, coherent identity, similarly the structure of the physical world is like unto a single being whose limbs and members are inseparably linked together." The Bahai approach is visible in the above stated para which encapsulates the oneness and wholeness of life, religion and people. Bahai faith necessitates for the respect of the Earth, of all creatures, preservation of Agriculture, restoration of the forests, balance and harmony of human and the environment, global and local co-operation as well as initiative is mandatory.

Baha'ullah through his teachings intended to develop a feeling of world citizenship and a sense of commitment to stewardship of the earth. His writings abound in deep reverence for the natural world and for the interconnectedness of all things. His words : "The earth is but one country, and mankind its citizens." Nature is reflection of the qualities and attributes of God. Bahai scriptures describe : "Nature in its essence is the embodiment of my name, the maker, the creature ... Nature is God's will and is its expression in and through the contingent world. It is a dispensation of providence ordained by the Ordainer, the All-wise." (Baha'u'llah, from 'Tablet of Wisdom', published in Tablets of Baha'u'llah revealed after the Kitab-i-Agdas (revised edition) Haifa : Bahai world centre, 1982). It teaches to train the future generation since infancy to be infinitely tender and loving to nature, animals, forests and humanity at large.

Similarly Judaism believes that "Whoever is merciful to all creatures is a descendant of our ancestor Abrahams. (Bezoh 32 B). Jews have been referred as "Merciful people" in the sacred writings of Jewish religion. Jews are ordered to do whatever is prescribed by moral conscience without expectation of a reward, immediate personal benefit. The encounter of God and man in nature is conceived in this faith as a "seamless web with man as the leader and custodian of the natural world." The sun, the moon, the stars and the plants, flame and fauna were created with a determined place and function in the universe by God. They can't encroach upon one another. When the environment faces dangerous and detrimental forces, the extinction of various species, both plants and animals, it is Jewish responsibility to defend the whole nature and to make it their primary concern.

VII

Moving forward from theoretical perspective to present day environment scenario of the world an attempt is being made to analyze the major international aspects of environment and ecology. The second half of the twentieth century marks a substantial upsurge towards the environmental awareness. Political leaders, Industrialists, Academicians and Concerned citizens realized the risks and hazards of multitudes of international environmental challenges.

Global environmental issues are addressed as 'Global' because some environment problems are not specifically local in terms of its influence. Some of the most important and relevant issues which are affecting global environment to a large extent may be mentioned as-the emission of US CFCs (chlorofluoro carbons) contributing to the global problem of stratospheric ozone depletion irrespective of its originating source. Similarly the engendering of harmful gases like carbon dioxide also tends to become an issue of international concern. Likewise the overuse of "global commons" viz resources which are shared by all members of international community, like atmosphere, outer space, oceans, deep-sea bed etc. are matter of great apprehension for the world. These issues are of global attention for various reasons because emission of sulphur dioxide in one state will pass through to another state by the air and ultimately it will be the reason of acid rain.. Similarly the disposed off wastes into one enclosed or semi-enclosed sea will create troubles in all states. Furthermore unsustainable agricultural practices, soil degradation, soil erosion, deforestation, river pollution and problems caused by urbanization and industrial practices all these are global environmental issues which are more or less associated with broader political and socio-economic processes and are an integral part of global economy.

Holes in Ozone layer of atmosphere caused by fluoro-carbons, nitrous oxide emitted by the industrial units on earth destroy this atmospheric Ozone layer and cause the creation of holes in this protective covering of earth resulting in ultraviolet Sun light reaching earth which will be responsible in increase of skin cancer. Dangerous atmospheric gases and acid rain are the result of petroleum products and military ammunition which produce dangerous gases like sulphur dioxide and amonia. Acid rain on earth is precipitated by these gases which contain sulphuric acid and nitric acid amongst other dangerous acids. The acid rain causes defoliation and death of trees, pollution of lakes killing all living creatures and ultimately infiltration of dangerous chemicals into the soil and ground water. Dangerous nuclear radiations are the result of the use of nuclear energy on a vast scale by the developed countries for peaceful and military purposes and it has put the environment on the earth in a danger of pollution by nuclear radiation. These all are the reasons for global warming and the warming up of planet earth too. Apart from these major problems, there are some prospective alarming issues also such as scarcity of fresh water, the decreasing land : people ratio, the emerging volatility of monsoon and erratic weather events, the slow pace melting of ice caps , rise in sea level, the birth of nonexistent diseases, loss of biodiversity and decreasing forest areas etc. Thus all these major issues need to be analyzed with their distinct characteristics and demand a separate solution oriented study, examination and subsequent conceptual renewal with follow up execution.

VIII

A retrospective study beginning from Stockholm Conference to Rio Earth summit onwards makes it evident that dozens of international environmental agreements, treaties, protocols and conventions were settled and executed from time to time with the objective of setting a transnational framework to approach the issue properly. Major international endeavours to safeguard the environment of the world are – Stockholm Conference on Human environment 1972, United Nation's Conference on water, Argentina 1977, Nairobi Declaration 1982,United Nation's Special Committee on environment 1984, Rio Earth Summit, 1992, (UN Conference on Environment and Development) UN Earth Summit, 1997, New York (to evaluate the implementation of recommendations made in Rio Earth summit), Quoto Protocol(International conference on Global warming) 1997, Johannesburg Summit on sustainable development, 2002 etc.

The Stockholm conference proved an icon in the development of international environmental policies. Being the United Nation's first major conference, a declaration containing 26 principles concerning the environment and development; an Action Plan with 109 recommendations spanning six broad areas (human settlements, natural resource management, pollution, educational and the environment, and international organization); and a Resolution on various institutional and financial arrangements were made. UNEP (formed in the conference) played a constructive role as a catalyst for many international environmental agreements and conventions.

The London Dumping convention of 1972 for restricting the dumping of toxic wastes (including nuclear wastes) at sea, the international convention of 1973 was to prevent international oil pollution from ships which was further strengthened in 1978, The Long Range Transboundary Air Pollution (LRTAP) agreement to limit emissions of sulphur dioxide and other pollutants causing air pollution and acid rain was set up by European and North American countries in 1979. The Vienna Convention for the Protection of the Ozone Layer was signed in 1985, which was followed two years later by the Montreal Protocol which imposed substantial limits on the use of CFCs and other ozone depleting substances.

Besides the organized efforts of United Nation and political institutions of different Nations, different environment NGOs came to command sufficient expertise and resources that they became substantial forces in international politics in their own right. Delegations from organizations such as Greenpeace, World Wildlife Fund, or Friends of the Earth at international meetings were frequently larger and more expert than those of all but the largest states through their access to the media and expertise were able to shape international agendas. Green movements and Green parties developed in most of the parts of world, and increasingly in developing countries too. In the USSR and Eastern Europe, environmental concern grew strongly in the 1980s. Environmental movements were at the forefront of broader political changes in these countries, and were particularly influential in most of the early post-communist governments.

The development agenda included in the Action Plan and Declaration of Principles agreed at Stockholm were never seriously followed up. Most of the international agreements focused on environmental protection or pollution without seriously integrating development concerns. As a result of this United Nation General Assembly decided to convene an 'Earth Summit' in Rio in 1992 in order to promote and develop sustainable development.

The Rio Declaration, Agenda 21, and the Declaration of Forest Principles were all agreed, and the conventions on climate change and biodiversity were respectively signed by 154 and 150 governments. The Convention on Desertification was not ready in time, and was not agreed until June 1994. Nevertheless, it is customarily included amongst the Earth Summit agreements. In the whole discourse of environmental debate North-South conflict is also an important factor in dealing with environmental issues globally and despite plans , programmes and policies for eco-protection the situation is getting worse.

IX

In nineties with the advent of globalization, a new debate started around the concept of globalization whether it is pro-environment or anti-environment. Many environmentalists in India and around the Globe consider it anti-environmental to a great extent. To quote Vandana Shiva, "We are so wrong to be smug about the new global economy ... it is time to stop and think about the impact of globalization on the lives of ordinary people." These environmental activists express their fear by putting forth some strong reasons as they observe that the disposal of waste of MNCs which are dumped into third world countries, who don't have adequate environment pollution and degradation control infrastructure and resources. Continually on the rise-an average of 220 tons of computers and other e-waste such as the outmoded models of electronic items (e.g. computers, t.v., cellphones, discarded and obsolete components and accessories) are not environmental friendly as these products contain highly toxic chemicals which pollute the soil, air and water.

While describing the impact of globalization on agriculture ecoactivists opinion that in the name of globalization of agriculture and development, seed companies are destroying the biodiversity of the land and depriving subsistence farmers of their livelihood e.g. in Warangal, seed companies lured pulses, millets and paddy farmers to buy hybrid cotton seeds named "White gold". The farmers now purchase "genetically engineered seeds" which need to be purchased every year at high cost and these hybrid seeds are too vulnerable to be saved and also ecologically harmful. The continuous sowing of the seeds renders the land infertile and hits the farmers economically. Thus genetically engineered seeds and globalization of trade in agriculture are creating hunger not feeding the poor.

The rich diversity and sustainable systems of food production are being destroyed in the name of increasing food production by monoculture farming. Planting only one crop in the entire field is monoculture and planting multiple crops in a mixture depending on weather or season is biodiversity. Monoculture will increase individual yield but planting various crops will give low yields of single crops but give high total output of food. As Vandana Shiva remarks "From the biodiversity perspective, biodiversity based productivity is higher than monoculture productivity. I call this(monoculture farming) blindness to the high productivity of biodiversity a 'Monoculture of the mind', which creates monocultures in our fields and in our world." Further, she strongly said that "economic globalization has become a war against the nature and the poor" and suggested to "move from market totalitarianism to an earth democracy", the survival of the human beings depends on the survival of all species and on the rules of the biosphere.

X

Coming to the Indian context, some of the major environmental problems caused by various determinants and pollutants are really pernicious to the present and future ecosystem of the country. Air pollution, water pollution and water scarcity, noise pollution, land degradation and soil erosion, deforestation, health and sanitation, rapid urbanization and population explosion- all these are major obstacles in India either hindering the economic development in rural area or straining the limits of services and facilities provided by the government in urban area, thereby placing significant pressure on natural resources and infrastructure of the largest democracy of the world.

The pollution of rivers, lakes and water ponds by industrial pollutants, sewerage, insincere human acts and deeds etc. are seriously affecting the environment and has been the major cause of water born deadly diseases. The scarce availability of water in regions like Rajasthan, Gujarat, Maharastra, Andhra Pradesh, Karnataka, Delhi and other states causes great hardship especially to the poor and women. The ongoing tussle on the mutual share of river water among states shows the drastic side of water problem. Cauveri water dispute between Karnataka and Tamilnadu, Haryana and Delhi over the water of Yamuna or the Satlaj between Punjab and Haryana are a few glaring instances of inter-state water-disputes.

Similarly air pollution in India, caused by several sources such as vehicular emissions, untreated industrial smoke, coal-fired power plants for electricity generation, carbon emissions, and rapid industrialization and urbanization etc. , is also one of the severe problems. Vehicle ownership is rising along with population and income. Although some reformative directives and measures by the government and the court have been taken from time to time such as setting the emission limit for gasoline and diesel powered vehicles at levels equivalent to Euro I and Euro II standards, prohibition on the use of 15 year old vehicle, the decrease in the motor fuel sulfur content, the ploying of CNG-fueled buses for commuters etc. Despite all these efforts air pollution is growing acute which makes metro life substandard. Increasing incidence of respiratory diseases such as asthma and bronchitis is recorded in cities like Delhi and Ahmedabad. Noise pollution caused by automobiles, vehicle horns, generators are on a high rise especially in big cities, near settlements on the main roads, and cross roads, bus depots and airports. Ambient noise levels in different commercial and traffic areas are far in access of permissible standards laid down for noise levels.

Land degradation is another major environmental problem and it has affected agricultural, non-agricultural, forests and land areas badly. Industrial pollution is caused by different industries such as fertilizer, pulp and paper, chemical (including dye-stuff/ colour) petrochemical, and engineering industries which are the main culprits of air as well as water pollution. These industries have been found defaulters in meeting air/water standards. The Bhopal Gas Tragedy is a reminder of chemical havoc.

Population explosion and urbanization are pressuring the Indian government to fulfill the increasing demands of consumption of electricity, food and eatable items etc. Urbanization is causing the villagers to rush for cities for many motives thereby increasing city population which causes health risks and so on. This description of the major issues concerned with environment is just a précis outlining the whole story.

XI

To meet out the above mentioned challenges, the constitution of independent India, implemented in 1950 commits itself to protect the environment through Article 48 – A of the Directive Principles of State Policy, as it declares that "the state shall endeavour to protect and improve the environment and to safeguard the forests and wildlife of the country." Further Article 51-A ,42nd amendment to constitution in 1976, directs the citizens to protect the environment as their fundamental duty, the 73rd constitutional amendment act of 1992 on the Panchayati Raj in India, and the 74th amendment act 1992 are significant from the point of view of environmental protection and management.

After the Stockholm conference in 1972, the government of India paid special attention towards the issues and problems centered round environment. This resulted in the establishment of the National Council of Environment Planning & Co-ordination in 1972 at the Department of Science and Technology and subsequently followed by another empowered committee in 1980 to review the then existing legislative measures and administrative machinery working for environmental protection. This committee recommended a separate department of environment which was set-up in 1980 and subsequently upgraded to a full fledged Ministry of Environment and Forest in 1985 to serve as the main machinery in the administrative structure of the government of India for the planning, promotion and co-ordination of environment and forestry programmes.

The department of environment in the states, Central and State Pollution Control Boards, the Botanical and Zoological survey of India, the Forest Survey of India, the National River Conservation Authority (earlier known as central Ganga Authority) the National Afforestation and Eco-development Board, the Indian Council for

Forestry Research and Education, the Wild Life Institute of India, the National Museum for Natural History, etc. are the units carrying out environmental activities in India.

To protect the environment major legislations in India are- the Wild Life (Protection) Act 1972, the Forest (conservation) Act 1980, the Water (Prevention and Control of Pollution) Act 1974, the Water (Cess) Act 1977, the National Environment Appellate Authority Act 1977, the Air (Prevention and Control of Pollution) Act 1981, the Environment (Protection) Act 1986, the Public Liability Insurance Act 1991 and the National Environment Tribunal Act 1995 etc.

The Ministry of Environment and Forests functions as a nodal agency for United Nations Environment Programme (UNEP), South Asia Co-operation Environment Programme (SACEP) and International Centre for Integrated Mountain and Development (ICIMOD), International Union for Conservation of Nature and Natural Resources (IUCN) and various international agencies, regional bodies and multilateral institutions.

To protect and safeguard the Environment India has also signed several important international treaties/agreements, just as (a) International Convention for the regulation of Whaling; (b) International Plant Protection Convention; (c) The Antarctic Treaty; (d) Convention on Wetlands of international importance; (e) International Convention on International trade in endangered species of wild flora and fauna; (f) Protocol of 1978 relating to the international convention for the prevention of pollution from ships; (g) Vienna Convention for the protection of the ozone layer; (h) Convention on Migratory Species (i) Basel Convention on trans-boundary movement of hazardous substances; (j) Framework Convention on climate change; (k) Convention on conservation of bio-diversity; (l) Montreal Protocol on the substances that deplete the ozone layer and; (m) International Convention for Combating Desertification etc.

India is also a party to the seminal document "Rio Declaration on Environment and Development" and also "The Agenda 21" both adopted at the United Nations Conference on Environment and Development (UNCED) at Rio de Janeiro in June 1992 in which more than 175 Nations participated.

In practice the policies, programmes and laws framed and executed by the government of India to meet the challenges of environmental degradation are violated by the interests of the profit making industries and business communities, manipulated by the politicians, bureaucrats to fulfill their vested interest. Poverty, illiteracy, ignorance, lack of environmental consciousness etc are some of the other socio-economic reasons for the disobedience of law.

The judiciary in India has played a constructive role to protect healthy life and survival of human beings. The Judicial activism of the eighties is remarkable in widening the scope and ambit of the right to live. The Judicial activism in response of environmental litigation in India symbolizes the anxiety of courts in finding out appropriate solutions for environmental pollution. To Quote Justice B. N. Kripal:

"The Supreme Court has recognized that the right to life guaranteed under Article 21 of the Constitution includes the right to wholesome environment... life, public health and ecology have priority over employment and loss of revenue problem...."

"While foreign investment in industry is being welcomed in India the entrepreneurs will have to bear in mind that environmental degradation will be resisted by vigilant public and therefore further industrialization should be ecofriendly."

XII

Eco movement is launched against the misuse and overuse of natural resources so as to sustain and restore the natural ambience of environment. A wide range of environmental movements against bioethical norms across the globe testify to the grim situation of political organizations. The reluctant approach of various political and governmental bodies towards the changing ecosystem causing threat to future ecology gives rise to environmental movement which is a collaborative social agitation consciously directed towards challenging the policies of the establishment. Therefore The initiators of such mass agitation are ecosystem people and ecological refugees.

The term ecosystem people denotes the category of impoverished people who depend on nature for living and livelihood. They fulfill their material needs from the natural environment of their own habitat. Due to the pressures of modern industries, technological breakthroughs and material growth natural world is deteriorating and nature's capacity to sustain its inhabitants who depend on it is shrinking. The upsurge of dams, industries, mining prohibits the local systems to fulfill basic needs of people. This situation displaces millions of forest dwellers, tribal peasants. The scarcity of water, recession of natural resources, recurrent droughts and famine compel people to leave forests. These people are categorised as ecological refugees or ecorefugees who live on margin of islands of prosperity such as sugarcane harvesters in South Gujarat and western Maharastra. Fortunate are big landowners, modern entrepreneurs, urban professionals, government employees who have purchasing power to buy consumer goods and can manipulate the state for subsidies, dearness allowances and raw materials, it is the ecosystem people and ecorefugees who are the victims in this process and consequently they initiate the environmental movement. They raise their voice against the negative impact of industrial development and growth oriented matters. Some major movements for the protection of environment are Chipko Movement1980, Silent Valley1980, Appiko Movement1983, Anti-Tehri Dam protest1991, Anti Vishnuprayag Dam protest11980, Cogentrix case1901-1996,Chilika Movement 1991. A brief outline of some ecomovements is mentioned below:

In 1980s, a 200 MW hydroelectric dam on the crystal clear river kunthipuzha under the kundremuckh project was proposed in Silent Valley in Kerala. But scientists, environmental activists and the local people opposed it as it was risk for certain endangered species and rare flower plants. Observing the seriousness of the situation Mrs. Indira Gandhi the then Prime Minister of India stopped this ambitions project.

The Chipko Movement became an 'ecological epic' in the Garhwal region, forced people to think that the main function of the Himalayan forest was to manage water not to yield timber. Appiko Movement of Karnataka was to stop obtaining timber from forests for commercial purposes. The Bhopal gas tragedy was the most tragic episode of the poisoning of the environment in human history with international dimension. It was an ecocide caused in India by a multinational company Union Carbide Corporation. It has raised the issues of the environmental prudence, human rights issues and MNC-Indian state relationship.

Tehri Dam being built on the river Bhagirathi in the Garhwal region of the Himalayas, partly funded by Russia is a major hydroelectric project. It is opposed on the ground of seismic data projecting earthquake disaster and displacement of the people of Tehri town and the villages in this region. Vishnuprayag dam was an ambitious hydroelectric project near Vishuprayag. This dam was planned to be constructed at a site joining the river Alaknanda and Bhagirathi just close to the world renowned Valley of flowers. It was opposed on the grounds that the dam would adversely affect the fragile ecosystem of the region and seriously damage the delicate world of the Valley's exotic plants and some rare endangered species of flower plants. Prime Minister Indira Gandhi gave her decision against proceeding with the project.

Chilika, situated in Orissa, is the largest brackish water lagoon in Asia. Fishing has been the traditional occupation for living and livelihood of people in chilika region. Export oriented instincts of industrialist supported by political parties to augment foreign exchange reserves of the nation by the export of prawn gave a death blow to the livelihood of 50,000 fisherman and dependency of two lacks people. An intense struggle by fisher people under the banner of "Matsyasibi Mahasangh" (Fisherman Association) started to fight for their rights. Utkal University students joined this struggle and started "Chilika Bachao Andolan" which eventually turned out a successful ecosaving movement.

Narmada river project covering four major states of western India provides a most important care to study in terms of maturing of environmental movement and dynamics related to politics of development. It brought the ecodevelopment problems to such a high level of informed debate, emotive rhetorics, political mobilization and grass roots activism. Narmada Valley Project (NVP) with its two mega-projects–viz SSP (Sardar Sarowar Project) and NSP (Narmada Sagar project in MP) is the largest single river valley project, with the world's largest man-made lake, ever to be planned in India.

The anti project activists give various reasons such as the reservoir will submerge thousands of hectares of forests lands, millions of Villagers will be displaced etc. There have been heated discussions and debates between pro-dam activists and anti dam protesters. This included political leaders and noted environmentalist Medha Patkar, Baba Amte and other high profilers. Still debatable the project has carved out a niche in the history of development and forced people and the Indian state to rethink for a new model of eco development.

Environmental movements are not confined only to conflict against the state, governmental departments and the commercial interests. They have developed positive aspects such as conservation and preservation of natural resources, e.g. forestation and water resource management.Seeing the enormity of the present and future ecological crises, the need of the hour is not the environmental movement rather Environmental Renaissance

The encapsulation of the multidimensional environmental issue in the introductory chapter of the book leads to figure out that sustainable development, ecodemocracy, environmental right to get unpolluted air, clean water, healthy food should be of prime importance on the agenda of all national and international governmental organizations for making policies and programmes along with good governance to execute these policies honestly, efficiently, effectively and timely. Civil Society, Citizens' active participation and role of media will be a great help to fight against the challenges of environmental protection.

XIII

Now we will embark to set the salient points of the papers that form this volume:

Dr. Pushpa Tiwari's paper titled "Environment Issues as Reflected in the Classical Thought of India : Some Reflections" is a deep exploration of classical legacy and heritage of Vedas, Upnishads Manusmriti, Kautilaya's Arthashastra to dig out the underlying principle of philosophy and cosmology of Indian world view which envisions man and environment as a unitary whole.

Dr. Krishna Banerjee's contribution to this volume is a sensitive and thoughtful article "Environmental Philosophy : The Indian

Perspective." The inexhaustible mystery of the dynamic relation between individual and cosmos, self and environment, in Agamic Tradition, Vedic Tradition, Sanskrit literary tradition and English literature have been described in the paper.

The elegantly written article "Environmental Protection and Awareness in Ancient India" by Dr. Sanjeev Kumar is a compact focused attempt to reveal ancient Scriptural notions about nature, ecology and environment and its constituents, forest and trees, sacred groves and gardens, wildlife and animals, water resources, water harvesting and conservation in view of man's relation with all these.

"Environmental Ethics", an enlightening article by Dr. A. N. Tripathi analyses how modern materialistic and consumerist attitude towards life, utilitarian view about nature's economic worth and techno-economic forces brought in their wake ecological imbalances compelling the environmentalists to talk about environmental ethics-what is our moral responsibility towards nature, ethical accountability towards future generations?

Dr. Suman Jain and Dr. Niharika Lal in their article "Environment in Contemporary Hindi Poetry" explore environmental natures and tradition in the light of Hindi literacy work.

The objective of Dr. Sunil Kumar Singh's paper "Environmental Education and its Concern in Education Policies in Independent India" talks to arouse environmental consciousness towards environmental crisis by introducing environmental education (EE). The paper also traces the development of environmental education at the international level and national level.

Dr. Priyankar Upadhyaya's article "Environmentalism – A Global Concern" traces global environmental problems and their reformative measures.

Dr. Rachana Srivastava's article "Environmental Issues – A Social Scientist's Concerns" views environment as an important subsystem of this broad social system, therefore social scientists who are concerned with social environment must sensitize

Legislature, Executive, Judiciary and Bureaucracy on the relationship between environmental issues and social engineering in a system.

Dr. Bhavana Trivedi in her article "The Political Dynamics of Environment in India" made an analytical study of policies and programmes of Environmental Politics in the Indian perspective.

Dr. Renu Srivastava's article "Social Challenges to Environment and Administration", unravels how society's fast pace hi-tech industrialization and urbanization process is jeopardizing the ecosystem.The international organizations UN, WB, IMF and WTO need not only to prepare policies and measures but to work within the framework of environment otherwise all the conferences on water and environment protection can't save our planet from irreparable damage.

"Environmental Economics: Challenges and suggestions" paper traces the relationship between economic activity and the environment. How Economic welfare must coincide with environmental welfare. The article is penned by Dr. Anup Kumar Mishra and Prof. G. S. Kushwaha.

"Bureaucratic Politics in Environmental Control" by Arun Kumar Upadhyay suggests for the reformation and expansion of Bureaucracy's institutional interests so as to give environmental agencies a substantial autonomy, budget and political support to implement regulations relating to environment effectively.

Rajendra Bharti's article "Environment and Administrative Role" defines administration, its role and function in managing environment crises with the support of constitutional provisions and judicial pronouncements.

Prof. Chandrakala Padia's article "Environment, Gender and Development" explores the concept of feminist environmentalism and the class-gender effects of the processes of degradation and privatization of Nature's resources.

"Women and Ecological Problems in Uttaranchal" under the co-authorship of Dr. Girish Chandra Pande and Aparna Joshi deals with the ecological problems and women's contribution to it in the newly constituted state Uttaranchal.

Dr. Devendra Mohan's article "Municipal Solid Wastes : Some Important Issues" defines Municipal Solid Wastes, its chemical composition, its treatment plan and decomposition in details. His other article "Pollution Due To Biocides" classifies the biocides and traces their negative impacts on Environment, effects on Human Health and Some Remedial Measures.

The article "Managing Mismanagement of Environment: A Planned Approach" presented by Dr. Shardul Chaubey focuses upon the managerial techniques for sustainable development.

Prof. C.M. Jariwala's paper "Indian Environmental Law : The finished and Unfinished Agenda" as its title indicates seeks to explore the bright and dark sides of Indian Legislature and Judiciary in the enforcement of environmental law, provisions, policies and how far the environment protection task is finished and what remains unfinished.

"Environment: Judicial Activism & Development of Environmental Jurisprudence" by Arvind Kumar Srivastava analyses the active role of judiciary in promoting the environmental justice.

The article on "The Environmental Provisions in the Indian Constitution" contributed by Dr. Tarun Kumar throws light on the several provisions of the constitution of Independent India for safeguarding the environment in India.

Dr. Sangeeta Jain's article "WTO and Environment" analyses the positive role of WTO in the environmental protection.

Dr. Kalpana Chakravorti's article "SAARC and Environment" explores the programmes and contribution made by South Asian Association for Regional Cooperation to protect the environment globally.

Dr. Dharmendra Kumar Mishra's paper "Environment : Different Issues" shows that man in his journey of progressive development first used the natural resources and ultimately exploited them , which led to environmental crisis.

Prof. Rajeshwar Upadhyay's paper "Environmental Resources and Population" compares and contrasts the increasing population explosion and decreasing environmental resources.

Philosophical and Literary Perspective

1

Environmental Issues as Reflected in the Classical Thought of India

Some Reflections

Dr. Pushpa Tiwari*

All history is a history of interrelation between man and environment. Environmental issues have acquired central status in the recent debates on modernity and development. In academia environment studies have become an independent discipline which directly intersects both social sciences, management and life sciences. This reveals the concern of present day scholars and thinkers relating to ecology and environment. This chapter tries to situate the environment issues of India's classical tradition in the milieu of contemporary chaos generated by industrialisation, development and modernity with the aim of having a comparative framework of reference for environmental concerns and anxieties. Global warming, extinction of flora and fauna, changing and erratic weather patterns, high level of atmospheric and ground level pollution, depletion of ozon layer, high level pollution of water, dying rivers, deforestation, soil erosion and soil salinity, pesticide havoc on insects and birds essential to delicate balance of ecology and environment—these terms and concerns have acquired the status of cliches because of their use and abuse in media-hype. It has diverted the main issue of ecological balance by focusing upon the immediate issues created by political lobbies and ad-hoc policies

* Reader, Department of Ancient History, Culture and Archaeology, University of Allahabad, Allahabad.

adopted by industrial powers. Biotechnology and Human Genome Projects are cases in point. A careful critique of environment issues depicted in media will reveal this trait of adhocism and response to the immediate challenges without any vision of the future perspective. As the environment related problems grow so the ad-hoc responses through politico-economic forums, summits and organisations. It has created a vicious circle of selfish motives and profit-oriented policy-plans to cash upon the problem. The socio-political and economic division of the world into two groups: (1) developed and (2) developing has further vitiated the scenario. Terms like neo-colonialism, neo imperialism, multinational, consumption oriented trade and commerce have become common in contemporary environment issues.

In this situation fuelled by unipolar yet economically dichotomised world more and more people are returning to their 'roots' i.e. to their classical legacy and heritage to find holistic vision and approach to deal with nature (प्रकृति). India is one of them. It has got a strong intellectual tradition which is often described as the classical tradition. Though the term 'classical' has multivariate connotations and value-laden meanings which a social science scholar should take into account. Here the term classical thought is taken in its wider connotation meaning those texts and textual traditions which constitute the basis of diachronic classical tradition e.g. *Vedas, Vedanga, Manusmriti, Kautilaya's Arthashastra* and *Brihatsamhita* to cite few of them. Epics like *Ramayana* and *Mahabharata, Puranas* (18 major and 18 minor), *Shilpashastras (Manasara, Aparajita-Prichchha, Samaranganasutradhara* etc.), Buddhist and Jain texts and so called secular literature consisting of poetry, drama, stories and chrita literature (Ashvaghosha, Kalidasa, Bhavabhuti, Banabhatta etc.), cannons on poetics and rheoteric (Kshemendra, Rajashekhara etc.) — there exists a wide range of textual tradition which comes within the purview of classical thought of anicient India. It is not possible to deal with all of them in a small article like this. It would suffice to say that they all reflect a kind of unifying underlying pattern of philosophical and intellectual ideas and ideals, Metaphysical vision and Cosmic blueprint which runs like a thread through beads of diverse shapes and sizes revealing a world-view which could only be described as the essence of 'Indianness' which is both unique yet universal in its meaning and application. The objective of the chapter is to propose an alternative vision of progress and development which treats ecology and environment as an essential part of human existence i.e. something integral to and inherent in it instead of something lying without.

The Sanskrit/Hindi equivalent of the English terms environment is supposed to be (पर्यावरण; परि आवरण) which is nothing but the literal meaning of the English word coined to suit the modern need of bilingual understanding. But a careful perusal of Sanskrit texts and lexicons will tell that in Indian intellectual tradition physical and metaphysical phenomena were never understood as something lying outside of existence and consiciousness or a thing inhabited by humans or, a thing encircling/encompassing human existence—which is what the English term and its Sanskrit/Hindi equivalent imply. Ancient Indian epistemology, ontology, philosophy and other cognate branches of knowledge have used the terms like 'प्रकृति जगत्, सृष्टि, भौतिक, आधिभौतिक, पंचमहाभूत, पुद्गल, आकाश-पृथ्वी, ब्रह्माण्ड' etc. to denote Cosmos, World, Nature, Creation, Elements, Space etc. Maya or the creative/kinetic force is said to be all pervasive which encompasses everyone and all levels of existence, hence it is described as 'आवरण' or false sence of division from the reality of true existence (Amarakosha, (ed.) Satya Deva Mishra, 1972, Malaysia, 3.3.80, 2.1.3 1.4.21, 1.7.37, 2.8.18, 3.3.73, 3.6.20, 2.10.11; Sanskrit - Hindi kosha, (ed.) Vaman Shivaram Apte, 2001 ed.)

The above deliberation sets the direction for culling out facts relating to environment from India's classical tradition of thought. It also reveals the underlying principle of philosophy and cosmology of Indian world view where man is not isolated from its environment in modern sence of the term. The relationship between man and environment is that of part and whole, but part and whole constantly interchange their positions in different contexts and situations. Hence, man and environment display a unique relationship where each is the condition of the other. The famous saying of "यथा पिण्डे तथा ब्राह्माण्डे" beautifully explains the relationship between part and whole, between micro and macro levels of existence. Whole is within the part as part is the total sum of the whole. For this reason, we find a world view in *Rigveda* comprising of space, air, fire, water, earth, sun, moon, stars, clouds, mountains, rivers, forests, herbs, grass, creepers, majestic animals, small insects alongwith man made things like chariot, domestication of plants and animals, houses, villages, towns, cities, boats, ships, weapons, dress, ornaments etc. etc. Almost every *Sukta* of *Rigveda* is devoted to personified nature where environment as a whole is prayed upon by *Rishis* and *Grihasthas* to bestow not only prosperity but also overall well being and peace. The underlying vision of *Regvedic hymns* suggests a belied in a system of intimate connectivity where each part/ element/ form is dependent on the welfare of the other to gain prosperity and joy. Hymns devoted to Sun, Sky-Earth, Earth, Energy (Agni), Varuna (Water), Rivers and forests reveal this holistic vision of divine connectivity and interdependence. Man is not there only to exploit the nature, rather to realize his divinity through his understanding of the Nature. *Rigveda* tells us about the harmonious existence of man/ environment by suggesting an ecosystem where welfare of each depends on the other. Thus it situates the relationship of man and environment within the ethical framework coming from within and not from without. By doing so *Rigveda* also fixes the responsibility and accountability of man to environment (*Rigveda*: 1.23.19-20: 1.34: 1.102.2: 10140-150: 1.159-60: 1.164.8: 5.84: 6.70: 7.101-102: 10.18.10: 10.65: 10.81-82: 10.146: 10.190).

In Yajurveda Dirghatama Rsihi says:

ईशा वास्यमिदं सर्वं यत्किंच जगत्यां जगत्।
तेन त्यक्तेन भुंचजीथा मा गष्धः कस्यस्विद्धनम्।।
कुर्वन्नेवेह कर्माणि जिजीविशेच्छतं समाः।
एव त्वयि नान्यथेतोश्यस्यि न कर्म लिप्यते नरे।।

(*Yajur*: 4.1-2)

Three things are suggested here for the over arching welfare of man and environment:

1. Divine connectivity to generate respect for every form of living and non-living organisms.

2. Concept of '*Bhoga*'(consumption) based upon '*Tyaga*' (renunciation/non-attachment) to avoid exteme exploitation of natural resources born of greed and profit.

3. Concept of duty (निश्काम कर्म) without any attachment to profit to avoid conflict, coercion, disharmony and imbalance in the ecosystem (जगत्).

These three principles need proper place in today's blind concept of unilinear progress and development based on the mindless exploitation of environment without having any regard for its sustainability.

Upanishads further reinforce this vision of divine connectivity. The mythological lores connected with Cosmology and Cosmogony describe the origin of man and environment from the same cosmic source. Individual souls are nothing but the manifestation of Single Soul known as Brahman (ब्रह्मन) hence, intimately connected with each other. This philosophic connectivity demands regards and reverence to each and every form of manifestation irrespective of their shape and size. Thus all are put on the same level as that of an elephant; grass and huge trees are treated as the two sides of the same coin; more important is the identification of each and every organ of human body with appropriate correspondences in nature, thus creating a cosmography where Nature/Universe/Cosmos resides within human form and vice versa. For example, *Aitareya Upanishad* describes following Cosmology and Cosmography:

तम् अभ्यतपत् तस्यभितप्तस्य मुखं निराभिद्यत तथाण्डं मुखाद् वाग् वाचो अग्निः नासिके निरभिद्येतां नासिकाभ्यां प्राणः प्राणात् वायुः अक्षिणी निरभिद्येताम् अक्षिभ्यां चक्षशः चक्षुश आदित्यः कर्णौ निरभिद्येतां कर्णाभ्यां श्रोत्रं श्रोत्राद् दिषः त्वङ् निरभिद्यत त्वचो लोमानि लोमभ्य ओषधिवनस्पतयो ह्दयं निरभिद्यत हदयान्मनो मनसष्चन्द्रमा नाभिः निरभिद्यत नाभ्या अपानों अपानात् मृत्यु......

(*Aitareya up.* 1.1.4)

From various organs of the heated body of the cosmic Purusha fire, consciousness (*Prana*), sun, moon, directions, herbs, vegetation etc. were born. This separation was removed by the Supreme being by allowing these gods (sun, moon etc.) to permanently reside in the corresponding parts of human body:

अग्निर्वाग्भूत्वा मुखं प्राविशत् वायुः प्राणे भूत्वा नासिके प्राविशत् आदित्यष्चक्षुर्भूत्वाक्षिणी प्राविशत् दिषः श्रोतं भूत्वा कर्णौ प्राविशन् ओषधिवनस्पतयो लोमानि भूत्वा त्वचं प्रवशिन् चन्द्रमा मनो भूत्वा हष्दयं प्राविशन् मष्त्युः अपानो भूत्वा नाभिम् प्राविशत...

(Aitareya up. 1.2.4.)

➔ i.e. fire ➔ speech ➔ mouth, air ➔ life breath ➔ nose, sun ➔ eyes-sockets, directions ➔ hearing power ➔ ears, herbs and vegetation ➔ body hair ➔ skin, moon mind ➔ heart, death ➔ life breath ➔ naval = human body cosmos = creation.

The above cosmology is important from two points of view. Firstly it believes in divine connectivity and interrelationship. Secondly, it portrays man in nature and nature in man i.e. to say they are the two sides of the same coin. Only if the environment/ nature is approached with humane sensitivity it can be expected to function in harmony with the divine connectivity and linkages of the universe. The inverse of this belief is also held: if the progress and welfare of human society is based on the principles of harmony and divine connectivity, if the development is in every way perfect, there will be blissful perfection in the Universe as well. The two way traffic of mutual progress and welfare is a must for the survival of man and environment.

Vedic religion and philosophy emphasise upon the reverence of natural elements particularly air, fire, water and earth, Specific rituals were performed to show the indebtedness of human beings to these elements for their survival. Modern notions of scientific mindset and progress have played havoc on the fragile balance of Universe and ecosystem by taking away the feeling of reverence and mutual regard; whereas, Vedic world view beautifully blends science and mythology to exhort human beings never to forget that every cell of their body in itself in a cosmos; that, welfare of cosmos is their responsibility; that by hurting nature they will hurt their own body and soul. As the *Aitareya Upanishad* mythology reveals the intimate and intricate web of connectivity of various parts of human body with different elements and forms of nature, now is the proper time to reflect on the rising trend of diseases relating to water and air pollution and reinvent the scientific cosmography of vedic tradition to sustain our civilization.

Rigvedic Seers were great psychologists and thinkers. They have critically analysed the traits of greed and false notions of progress based only on material wealth and consumption. In one of the hymns they have given their psychoanalysis of human nature as follows:

एकपाद भूयो द्विपदी द्विचक्रमें द्विपाद् त्रिपादमभ्येति पश्चात्।

चतुष्पादेति द्विपदामिसरे संपश्यन्पक्तीस्पतिष्टमाना।।

(*Rig*: 10. 117.8)

The above mantra describes the human desire of aping those who possess more wealth and the vicious circle of never ending envy and blind process of running after material weatth. Vedic seers have put forward a philosophy of restrained consumption based upon unity and mutual respect for each others need. They pray to God to develop the holistic vision of friendship and regard for each other:

दृते दूँह मा मित्रस्य मा चक्षुष सर्वाणि भूतानि समीक्षन्ताम्।

मित्रस्याहं चक्षुषा सर्वाणि भूतानि समीक्षे मित्रस्य चक्षुषा समीक्षामहे।।

(*Yajur*: 36: 18)

Today's world of greed and exploitation needs the above mentioned holistic vision as the corrective measure. The Vedic concept of '*Vasudhaiva Kutumbakam*' is not limited only to the philosophy of global market, global village, global trade and commerce as is envisaged by the multinational big industrial establishments to achieve their targets of profit. Vedic vision also includes global harmony, global sensitivity and global unity and welfare. It firmly believes that by hurting one section of the globe one will be jeopardising the very existence of the globe and also whole of the Universe.

Thus, Vedic tradition does not make any distinction between the welfare of human beings and that of the Nature. Contrary to modern dichotomy of man and environment it treats both man and environment as two sides of the same coin, hence mutually inclusive of each other. Its philosophy of progress and development is based on the mechanism of restrained consumption (भोग + त्याग) and mutual regard (आत्मवत् सर्वभूतेषु) for each other.

Environment issues are intimately and directly related to political systems, policies and control mechanism. For this reason, I have chosen two great thinkers of polity and social customs belonging to the classical intellectual tradition of India. They are Kautilya (4-3rd cent. B.C.) and Manu (2-1st cent B.C.). Kautilya's *Arthashastra* in this sense is secular and pragmatic as it is concerned with the welfare of state and king. *Arthashastra* is a much maligned text in academia. Modern notion of dichotomy between matter and spirit comes into the way of scholastic understanding of this text in academia. Modern notion of dichotony between matter and spirit comes into the way of scholastic understanding of this text. As it is devoted on the elaboration of 'Artha' most of the scholars have interpreted it on criteria of Marxism, Capitalism, Monarchical and Autocratic systems etc. etc., thus resulting in one sided and narrow discourse of political economy. These interpretations have missed the wood for the trees. It is true that Kautilya has given primacy to '*Artha*' (अर्थ) over the other three ideals. But equally true is the fact that his vision of '*Artha*' is not mutually exclusive of the other ideals; it is so broad and flexible that it not only includes the importance of *Bharma* and *Karma* (धर्म, काम) but treats them in a cyclic notion of 'origin where one ideal gives birth to the other. This cyclic approach presupposes the contextual primacy of every ideal in the holistic vision of progress and development. For example Kautilya states that:

सखस्य मूलं धर्मः। धर्मस्य मूलमर्थः। धर्मस्य मूलमर्थः। अर्थस्य मूलं राज्यम् राज्यमूलमिन्द्रियविजयः। इन्द्रियजयस्य मूलं विनयः। विनयस्य मूलं वृद्धोपसेवा। वृद्धसेवाया विज्ञानम्। विज्ञानेनात्मानं सम्पादयेत। सम्पादितात्मा जितात्मा भवति। जितात्मा सर्वार्थैः संयुज्येत।

i.e. true joy originates from Dharma, Dharma from Artha, Artha from State, State from control over faculties, control of faculties comes from politeness, politeness comes from performing service to elders, this kind of service originates from true knowledge (विज्ञान); therefore, every human should strive for gaining true knowledge. A person of scientific temperament conquers himself i.e. the fickleness of mind and faculties. One who could exercise total control over the self could achieve '*Artha*' in its totality (सर्वार्थैः संयुज्येत) [Kautilyiya Arthashastram,(ed.) V. Gairola, Varanasi: Chowkhamba Vidyabhawan, 4th ed. 1991, p. 775]. Thus, the cycle begins and ends with joy (सुख) which is the aim of every human thought and action.

The very beginning of *Arthashastra* states the aim of the text: 'पृथिव्या लाभे पालने च' i.e. both exploitation and sustenance of the earth this knowledge is described in the *Arthashastra*. According to Kautilya Trayi (त्रयी) lays the foundation for understanding the holistic spiritual vision (धर्म-अधर्म), Varta (वार्ता) (agriculture, trade, commerce) is necessary to understand the scientific, pragnatic and non-disastrous methods (अर्थ-अनर्थ) and *Dandanity* (Polity) teaches about the relevance of good policies and management (सुशासन-दु:शासन). While describing the life style of a sovereign king Kautilya says :

एवं वश्येन्द्रिह : परस्त्रीद्रव्यहिंसाश्च वर्जयेत्।

...... अधर्मसंयुक्तमानर्थसंयुक्तं च व्यवहारम्।

(*Arthashastra*: 1.3.6)

i.e. ruler should exercise total control over his faculties. He should not desire others women , property / money; he should not indulge in violent policies and methods; he should discard unlawful and disastrous policies and practices. The analysis of *Arthashastra* discloses the milieu and framework within which Kautilya was unfolding his discourse. Following essential features of the discourse are discenible in the text :

1. Restraint over collection and consumption;
2. Proper management of natural resources;
3. Welfare of state and management of natural resources;
4. Knowledge of agriculture, horticulture, husbandry, water resources, forestry, mining necessary for overall prososperity;
5. Sustainable development and production policies to be followed by the state;
6. Mutual respect and regard for interdependence.

All these features constitute the eco friendly political economy of *Arthashastra*. Kautilya has emphasised the existence of green belt and water resources near human establishemnt of a village, town or district:

नदी शैलवनगृष्टिदरीसेतुबन्धशाल्मलीशमीक्षीरवृक्षानन्तेषु सीम्नां स्थापयेत् ।

i.e. river, hills, forest, artificial lakes and reservoires alongwith tropical trees like Semala, Shami, Banyan etc. should form the dividing line or border between two habitational sites (2.17.1). Kautilya advocates intergrated development schemes and knowledge. For example, according to him an authority dealing with agriculture should also have the knowledge of agro-technology, forestry, botany as well as the art of mathematical measurements (2.40.24). Authority concerned with mining should have the knowledge of measurement units metallurgy, chemistry, furnace-technology and heating the chemicals, geology and gemology (2.28.12). In fact the whole scheme of the Second section (*Adhikarana*) of *Arthashastra* is based on the principle of integrated sustainable development which should be adopted by the state. He exhorts kings and state officials to implement the scheme of converting the barren, infertile lands into fertile agricultural land (2.18.2). Restraint and sustainability, exploitation and preservation these form the basis of Kautilya's *Arthashastra*. Contemporary environmental issues need this philosophy of restrained consumption and intergrated approach of development if they are aimed to have an effective control over mindless exploitation and destruction of the globe.

Another classical thinker of India is Manu. *Manusmriti* gives us the detailed account of socio-religious customs and beliefs and the ideals of human life according to the bio-cultural needs of varrious categories. Three things emerge from the critical analysis of this text. Firstly, the sense of divine connectivity underlying this world and whole of the Universe thus connecting the smallest/ lowest of matter with the highest form of manifestation :

सर्वभूतेशु चात्मानं सर्वभूतानि चात्मनि ।

(*Manu*: 12.91)

Secondly, the ideals of family life [गृहस्थाश्रम] based on restraint and control over, needs :

न जातु कामः कामानामुपभोगेन शाम्यति ।
हविशा कष्ष्णवर्त्मेव भूय एवाभिवर्धते ।।

(*Manu*: 2 . 94)

Mere consumption and indulgence can not satisfy need and desire; rather, unrestrained indulgence aggravates and intensifies desire the way purified butter intensifies the fire.

As a corrective measure *Manu* suggests, the golden principle of indulgence and restraint (भोग-त्याग), Knowledge and wisdom (तप-विद्या) to sustain harmony, peace and prosperity:

यश्चैतान्प्राप्नुयात्सर्वान्यश्चैतान्केवलांस्त्यजेत् ।

प्रापणात्सर्वकामानां परित्यागो विशिष्यते ।।

(*Manu*: 2 . 95)

i.e. one who indulges in all kinds of consumption is inferior from those who renounce indulgence; thus renunciation is held superior to indulgence and consumption. Thirdly, ideals and values of humanity and individual as :

धृतिः क्षमा दमाऽस्तेयं शोचमिन्द्रयनिग्रह : ।

धीर्विद्या सत्यमक्रोधो दशकं धर्मलक्षणम् ।।

(*Manu*: 2 . 92)

A careful perusal of Manusmriti reveals the scheme of integrated growth and development underlying its discourse of *dharma, Artha, Kama* and *Moksha* (धर्म, अर्थ, काम, मोक्ष), Purity and preservation of earth, water, air and balanced human attitude towards nature given primacy in the 'manavadharma' (religion of man). Rituals, ceremonies and sacraments were used to protect the nature and also to make human beings accountable to its welfare by identifying human body with the universe/Nature through cosmography.

6-7 century A.D. *Brihatsamhita* of Varahamihira is a veritable storehouse of contemporary knowledge, weather prediction, science and technology of agriculture, water conservation tree and vegetation, animals, architecture, textile etc. etc. Environment issues are expressed in almost all the chapters of this text. Particular attention has been given to agriculture, metals, precious stones, water, trees etc. by devoting each of them a separate chapter. What is more important is the belief of astronomical impact on agriculture, water etc. which reflects the practical side of Vedic cosmography and cosmology and its principle of divine connectivity. Its chapters like 40 'सस्यजातकाध्याय' (agriculture), 41 'द्रव्यनिश्चयाध्याय' (metal, stones and other precious items), 53 - 'वास्तुविद्याध्याय:' (Architecture and planning), 54 - 'वृक्षायुर्वेदाध्याय:' (underground water), 55 - '' (Science of trees and forestry) etc. throw significant light on the environmental issues as conceived and perceived by the society of the time of Varahamihira. A careful and critical analysis reveals a world view of divine connectivity and mutual interdependence of man and environment. The simple technique of identifying underground water and its sustenance is an eye-opener. Modern science has given the tool of mindless exploitation and destruction of nature to human beings. Without having any regard for the purity, conservation and preservation of life giving elements like water and air, policy of mass scale industrialization growth is being followed in the name of progress. It has put a big question mark before the existence of human civilization itself. Like a vicious circle every policy of correcting the environment generates further chain of destruction and pollution. For example building big dams, cutting forests, green revolution, canal system, urban planning, mining etc. they all have generated the danger of dying rivers, soil erosion, salinity, pesticide related pollution, concrete jungles and polluted atmosphere, depletion of minerals etc. This happens because of the isolationist vision of tightly compartmentalized policies of development which does not accept the principle of connectivity and totally negates mutual regard for the other sectors. This is an irony of present day science that on the one hand it reveals the single underlying mechanism of unity and on the other practices extreme isolationistic policy of development through technology. It emphasises the multi and cross disciplinary approach but allows narrow vision of exploiting natural resources without any future concern. Relevance of *Brihatsamhita* seems to be in this area of bridging the gap between theory and practice. While elaborating the technique of underground water it emphasises the role of trees and grass to realise the knowledge of water level. By this simple method it also paves the way for water preservation by protecting related vegetation and trees. In the same way while describing the science of trees it throws significant light on soil preservation. Science of architecture is described as the science of space and time in relation to human and physical geography. Architectural planning replicates the celestial movements and cosmography. Thus, knowledge of soil, preservation of trees and water resource are essential to architectural planning construction of house, city. Conglomeration of buildings without any regard to its environment. Trees, ponds, wells, lakes, groves constitute an integral part of the whole planning. This is the essence of the futuristic policy of any sustainable development programme which now policy makers are trying to emphasise to stop the destruction of earth and its life giving/life sustaining environment.

Thus, this bried sketch of the environment issues as contained in the classical thought of ancient India emphasises the need of appropriating traditional science and technology to the needs of modern circumstances so as to stop the monster of industrialisation and mindless consumption of natural resources. It also explains the holistic vision of ancient thinkers had their philosophy of restraint and indulgence. In *Matsyapurana* Goddes Parvati has said that:

> एवं निरूदके देशे यः कारयेद्बुधः
>
> बिन्दौ बिन्दौ च तोयस्य वसेत्संवत्सरं दिवि ।।
>
> दशकूप समावापी दशवापीसमो हृदः ।
>
> एषैव मम मर्यादा नियता लोकभाविनी ।।

(*Matsya Pu*: *Purva ghage* Chap. 154.511-512)

i.e.'one who constructs a well in an arid area he enjoys heaven in proportion to every drop of that well. One pond is equivalent to 10 wells, one lake is equivalent to 10 ponds; One son is equivalent to 10 lakes, and one tree is equal to 10 sons. This is the ideal way of life which I have established.' The system of pilgrimage was introduced to harness the energy of religion, tourism and trade. *Shilpashstras, Puranas* and *Agamas* exhort human beings to construct wells, ponds, lakes and plant trees on the site of temple construction and pilgrimage. This eco-friendly system of progress and development needs reinvention and revitalisation. We could conclude that we need the synergy of mutual interaction between religion and philosophy, and socio-economic and political planning. The environment issues of India's classical thought tell us to discard the isolationistic vision of progress where one segment flourishes at the cost of other. It also emphasises the need of bridging the gap between matter and spirit, between knowledge and wisdom, skill and expertise and more than anything else between progress and existence. We require the holistic vision of Vedic tradition, rationality of Kautilya and ingenuity of Varahamihira.

2

Environmental Philosophy

The Indian Perspective

Prof. Krishna Banerjee*

In the integral, holistic lived Philosophy of India, environment is not the other, the external, to be conquered, tamed and annexed for selfish purposes. The self and environment are two points of reference, dynamically related in mutual feed back.

In a sense, the self is the focal point of environment, and environment is the irradiation of the self. *"Otaṃca Protaṃca"* is a recurrent phrase in th *Upaniṣad* indicating the mutual interaction of the microcosm and the macrocosm the Absolute and the relative/relational.[1] The entire *Upaniṣadic* metaphysics is devoted to various descriptions of the relative as related to the Absolute, mapping out in metaphors, myths and cryptic phrases as to how the universe has spheres within spheres, systems within systems, of interrelatedness.

The *Āgamic* tradition corroborates the same idea in colourful delail.[2] The *Tantra Śāstra,* for example, describes the nerves of energy currents in the body as rivers, the spinal column as the Mount Meru, and the three principal energy currents, *Ida, Pingata* and *Susumna* as the three sacred rivers, *Gangā, Yamunā* and *Sarasvatī*. In fact, this parallelism between *piṇḍa* (The body construct) and

* Professor of English, Women's College, Banaras Hindu University, Varanasi.

brāhmaṇḍa (the cosmos) is fundamental to the systems of *Jyotiṣa* (astrology), *ayurveda* (medicine), *Vastukalā* (architecture), *murtikalā* (sculpture), *citrakalā* (painting), *sāhitya* (literature) and *sangīta* (music).

In order to understand an idea in the context of the Indian tradition, one has to see how it appears in the revealed Word chanted by the seers (*śruti*), in contemplation on and interpretation of the revealed Word recorded by the sages (*smṛti* and *darśana*), in individual and social life (*āearā*), in the cultural heritage such as literature and art (*sāhitya, kalā*), and in codes of ethics (*nīti* or *dharma*). One feels the pulsating presence of environmental awareness in all these dimensions of life in India. Its all-pervasiveness has often been grossly misconstrued. Vedic hymns, for example, have been regarded as pantheistic and polytheistic worship of natural phenomena. Vedic scholars have tried to clear up this misconception. Sri Aurobindo, for instance, writes in *The Secret of the Veda* that the Vedic gods, *Agni, Sūrya, Indra, Varuṇa,* and so on, are subtle psychic and spiritual forces working within the body-mind-spirit complex of the individual (microcosm), that by propitiating these forces, one progresses on the path of ascension to the highest spiritual goal (Oversoul).[3]

Even as inspired poetry, the hymns of the *Vedas* are exquisite expressions of reverence, awe, love and wonder experienced by sensitive minds in close contact with nature. As Dr. S. Radhakrishnan observes, the Vedic hymns are "the outpourings of peotic minds who were struck by the immensity of the universe and the inexhaustible mystery of life.[4] The following utterances of the Vedic seers are just a few examples of how natural environment was near and dear to our ancestors.[5]

"Live in complete harmony with nature,

experience the grace of God in the

Splendour of the universe." (*Yajur Veda* 34.37)

"Whatever I dig from thee, O Earth,

may that have quick growth again.

O purifier, may we not injure thy vitals or thy heart." (*Atharva Veda,* 12.1.35)

"O Mother Earth,
Let Thy bossom be fee
From sickness and decay,
May we through long life
Be active and vigilant
And serve Thee with devotion." (*Ath. V.* 12.1.62)

"Waters contain
All disease-dispelling medicaments,
Usefull for the upkeep of our body,
So that me may live long
To enjoy the bright Sun." (*Rg. Veda* 1.23.2)

"The waters are the sovereign remedy,
Herbs possess divine efficacy," (*Ath. V.* 8.7.3)

"From time eternal, the Mother Earth is
giving life to her children—you owe
debt to Her." (*Ath. V.*, 12.1.45)

"Bless that the sun,
with extensive radiance,
Rises for peace.
May the four quarters of the horizon
Be auspicious for peace and harmony. (*Rg. V.*, 7.35.8)

"Ye cows, You fatten the emaciated,
and you make the unlovely beautiful,
Make our house happy, you with pleasant lowing,
Your power is glorified in our assemblies!" (*Rg. V.*, 6.28.6)

"May you have many calves, graze on pastures
and drink pure water at drinking places,
May not the thief master you, nor the wicked,

And may the darts of the fierce Lord leave you aside." (*Rg. V.* 6.28.7)

The *Upaniṣads* have specific chapters on cosmic genesis and cosmological classification, which are like a rhythmic dance of unification and diversification, contraction and expansion. Like the famous *Puruṣa Sūkta* of the *Veda,* the opening chapter of the *Bṛhadaraṇyaka Upaniṣad,* which metaphorically describes the world as a sacrificial horse,[6] suggests that the cosmic creation is not a composition of multiple components but the gradual diversification of a basic unity. The diversified strata in nature are so interrelated that disharmony in any one of them disturbs the harmonious function in all the rest. It should be the duty of each responsible person to see that depletion in nature is followed by compensation and pollution by proper cleansing.

Yajña or sacrifice is one of the key-concepts relating to the dynamic stability of the cosmos. The physical universe as the body of the *Virāt* (the Great one refreshes and reconstructs itself through self-sacrifice. The universe has come into existence through *yajña,* and each human being shares the responsibility of maintaining environmental sanity and the cycle of renovation in nature by participating in *yajña.* Each individual is indebted to society, to nature and ultimately to the Creator for all that he/she has got, including his/her very existence. It would not only be an act of ingratitude not to extend one's helping hand to society and the natural environment but it would also hamper the cyclic process of give and take that sustains life's dynamic equilibrium. Thus the scripters recommend daily performance of *yajña* for all adult persons excepting the sannyasins, the renunciates.[7]

Yajña is the reverential offering of part of one's possessions, made for promoting the well-being of all. All things we get from life are to be shared with others, for we are indebted to the entire environment for whaterver we enjoy. This Vedic idea of *Yajña* is succinetly dilated upon by the Lord *Kṛṣṇa* in the *Bhagavad Gīta,* with additional shades of meaning.[8] Infact, *Karmayoga,* the central message of the *Gita,* is the philosophy of action performed in the spirit of sacrifice. As agents, we may be selfishly arduous, foolishly negligent or wirely oriented to the secret of action. The Creator has initiated the cycle of Karma, sustained by nature. This great chain of cause and effect forms a rotating circle. Viewed in this perspective, even a small individual act has a long past of causative factors and an equally vast future of possible effects. When we place ourselves in this never-ending cycle, we feel humble, as our very agenthood seems to be illusory. With humility and grateful reverence, we should share with worthy recipients all the gifts that life and nature have bestowed upon us. One who cooks food only for oneself commits an act of theft; sin contaminates that food which is not shared by others.[9]

The *Manu-Smriti* talks about the *pañca-sūñā,* the five sins of the householders commited daily, though unintentionally, as small insects and other creatures are harmed by the use of the kitchen fire, the broom, the grinder, etc., and the *pañcā mahāyajña,* the five great sacrifices by which those sins get absolved. These are *Brahma-yajña* or daily study of scriptures as an offering of tribute to Brahma and also to the enlightened masters; *Pitṛyajña,* the offering of homage to the Manes; *Devayjña,* performance of *homa* by offering oblations to the *Devas; Bhauta-Yajña* or *Bhūta-Yajña* offering food, water etc. to animals and birds, and *Nṛyajna,* the giving of food and objects of utility to needy human beings, especially to the *atithis* or persons who arrive at one's door, seeking help.[10]

The concept of *prāṇa,* like that of *Yajña,* is pivotal to the *Vedic-Upaniṣadic* cosmology. Like fire, the vital breath unifies the physical with the metaphysical.[11] Much more than air, much more than breath, *prāña* is the life-principle that manifests in air, in breath, in vital energy and many other modes. Environmental sanity is conducive to the proper functioning of this subtle principle. In *chandogya* and *Brhadaranyaka Upanisads, prāṇa* is described as inviolable and as the supreme controller of the senses. The elaborate purificatory rites of the orthodox Hindus are based on the belief in pranic vibrations. Long before J.C.Bose demonstrated in his laboratory that plants have life and sensitivity, ancient Indian scriptures had already described them as living and sensate. The *Manusmrti* says that frees are *antaḥsanjña,* endowed with inner consciousness and the feelings of joy and sorrow, though they are unable to express them.[12] In India, elders teach children not to pluck flowers and leaves after midday nor at night, since it is believed that trees feel less pain in the morning than later. We are told that while collecting leaves from a tulasī plant, one should hold the branch softly with the left hand and pluck leaves carefully with the right hand, without causing damage to the twigs or the bark of the plant. Knut A. Jacobson, in his recently published book, *Prakriti* in *Sankhya-Yoga,* has analysed the concept of ahimsa in terms of environmental ethics.[13]

Religious rites and festivals celebrated the year round put emphasis on cleanliness, peace and harmony of the environment as well as underscore the importance of herbs, flowers, fruits, leaves, roots and powders collected from nature. This practice helps cultivation and preservation of useful plants. The special importance that certain animals like the cow have revceived is owing to the recognition of their value is human life. This recognition is above utilitarianism. It is based on love and the notion that all lives are intrinsically valuable. An example is the custom of releasing a live fish in water on one's birthday. Birds and animals have found a place in the pantheon as vehicles of the gods and goddesses. The colourful varieties of life in nature are thus sampled and preserved in the mythology to be remembered in moments of festive joy.

Our sacred plants are found to be specially rich in medicinal properties. Kaviraj Shibkali Bhattacharya in his book on herbal treatment *Chirnjiva Vanaushadhi* in 6 volumes, has meticulously traced the history of Indian medicinal herbs back to their description in the *Vedas,* highlighting their role in religious ceremonies and in social-cum-personal life.[14]

Our philosophy holds that the body is made up of *pañca-bhutas,* the five elements of nature, namely, earth, water, air, fire and ether, It logically follows that these elements that form our natural environment ought to be maintained in a good condition so that their benign influences may keep us hale and hearty.

The ritualistic worship of the deities is another proof of how much importance is attached to the environs. The *pañcopachāra,* the five offerings, relate to the five elements. The flame of the lamp represents fire; the incense relates to air; the food offering stands for the earth, of which it is a product; the water offering hallows the element of water, and flowers, Natures marvellous things of beauty, represent *ākāśha,* ether or space. This space is also *daharākāśa,* the space within the heart, which receives the joyous communication of beauty. Thus ritualistic materials, culled from nature's immense store house, help maintaining man's relationship with nature.

The peace-chants of the *upaniṣads* are replete with environmental awareness, as they are rich in poetic beauty.

"The winds blow sweetly for the righteous, the rivers pour forth honey. May the herbs be sweet unto us. May the night and the day be sweet. May the dust of the earth be sweet, May heaven our father, be sweet to us. May the tree be sweet unto us. May the sum be sweet, may the cows be sweet unto us."[15]

This prayer is thematically related to the Honey Doctrine, *Madhu-Vidyā,* which is one of the several mystic doctrines of the *Upanisadṣ.*[16]

Madhuvidyā establishes the fact that the cosmos and the individual are essentially interdependent as the same all-permeating self is present in both, effecting an inter supply of succour and sweetness between the cosmic and the idividual. "This earth is (like) honey for all creatures, and all creatures are (like) honey for this earth." Similarly, water, fire, air, men, the quarters, moon, lightining, cloud, sapce, law, truth, mankind and self are respectively called honey for all beings as all beings are honey for them.[17]

Prof. Joy Sen of I.I.T, Kharagpur believes that while *Brahmavidyā* teaches how one should prepare oneself for *mokṣa,* the total release from the shackles of ignorance, *Madhuvidyā* is an applied philosophy that teaches the art of harmonious living in art, architecture, medicine, aesthetics, etc., paradigms are drawn from nature, showing the inter-relatedness of the three parallel worlds, the spiritual, the mental and the material. Prof. Sen holds that *Madhu-Vidya* is an applied form of *Brahmavidyā*. In a television talk on DD7 (1.6.2003, 9.00-9.30 a.m.) he informed the viewers that at I.I.T., Jadavpur, research programmes have been undertaken in order to investigate into the relevance of *Madhuvidyā* to plans of ecofriendly constructions.

In Classical Indian poetics and classical Indian music, the two key-concepts of rasa (meaning both essence and joy) and rag (signifying colour as well as love) are based on the idea of reciprocity and compatibility between the inner atmosphere or psychic mood and the outer atmoshpere or natural environment.[18]

In conclusion, I refer to *Kãlidãsa,* who in his unique way has highlighted our ancestors' love of natural environement. The following verses from his epic poem *Raghuvaṃśam* present beautiful glimpses of *tapovana.* Having graduated from the *Gurukula,* the brahmin boy *Kautsa* visits the great king *Raghu* in his court with the interntion of requesting the king to provide him with the amount that he has to offer to his *Guru Varatantu* as *Gurudakṣinā. Raghu* welcomes him with due reverence and joy and makes inquiries about the well being of the hermits and their environment:

"I hope storms and other calamities have not injured the trees of the hermitage, which remove fatigue, and which have been fostered with the care bestowed upon a son, by constructing water-basins and such like operations.

Are the fawns of the deer safe—the fawns whose desire for the Kusha grass is indulged by the hermits, through their love for them, although the grass is required by them for religious rites, the fawns whose umbilical cords drop down in the hermits laps, which form, as it were, their beds.

Are the holy waters unsullied, by means of which the necessary ablutions are performed and the libations offered to the manes, and the sandy banks of which are marked by a sixth part of the gleanings of corn.

I hope the ripe *Nīvāra* corn and similar forest produce upon which you depend for subsistence and a part where of is to be set apart for guests who arrive in time, is not eaten up by country cattle that feed upon chaff. (*Raghuvaṃśam,* VI, Slokas 6-9)

When Wordsworth wrote:

"And I wish my days to be

Bound each to each by natural piety." (*The Rainbow",* Lines 8-9),

He expressed the wish that had been amply fulfilled in the *Tapovana* of India. In *Abhijñānasākuntalam,* Kālidāsa has depicted more colourfully how trees and creepers, birds and animals, brooks and hill were given reverence and love. Such ecological concern was not prompted just by economic motives, it was a concern for Nature as a living and loving entity.

"O lady! we receive but what we give,
And in our life alone does Nature live."

(*"Dejection:An Ode"*, lines 47-48)

Coleridge's brilliant insight expressed in the above lines has been tacitly accepted in India as a fact established through ages of philosophical contemplation and customs of daily life. The Indian *Tapovana*, the forest of the hermits, was neither a wilderness nor a cultivated fashionable garden. Without any greedy or callous interference with Nature's ways, man here lived in close companionship with Nature, enjoying a reciprocal bond of care and love.

Talking of *Śakuntalā* and her watering pot symbolizing love of Nature, the eminent orientalist, Henry W. Wells, writes :

"She seeks to modify nature only in so far as she seeks to bring warmth, encouragement and affection into her relation with all living things... . To the parched shrubs and trees she carries her watering pot. She makes no attempt to alter nature except in so far as she befriends it. Wholly accepting nature's plan, she imposes no programme or scheme upon it. The sacred grove in which she lives as a girl and young woman accordingly is and is not a garden. It is no garden in the sense that:

In it man plants nothing and arranges nothing.
But it is a garden in the sense that all that
lives and grows is encouraged with a loving hand."[19]

Wells further writes, "... the supreme tribute offered by the Hindu imagination to nature was the sacred grove. In such a grove, Krishna, best loved of gods, disported himself. In such groves gods themselves lived. And in such *Sakuntala* flourished, like the fairest tree of all. Her watering pot is symbol of her outlook. It is not excessive to declare that no highly civilised people has lived in such close spiritual intimacy with nature as the ancient Hindus."[20] "He adds "the Hindus are the true, glorious children of nature...".[21]

Referring to Indian idealism, Wells observes, "It gives us first of all the love of living things, as witnessed in sculpture that teems with vital forms. Hence, of course, derives the Hindu preference for a vegetable diet, plants being less vividly and crucially alive than bird or beast. Solicitude for the forms of nature is at the heart of the *Sakuntala* image."

This observation call to mind the unforgettable scenes of the play, as for example, when *Vaikhānasa* the hermit stops the king *Duṣyanta* in pursuit of the running deer, saying, "O King, this is a deer of the hermitage; he must not be slain, he must not be slain", or when *Śakuntalā* declares to her friends *Anasūyā* and *Priyamvadā* her love for the trees, saying, "It is not only is obedience to our father (that I take care of the trees), but I really feel the affection of a sister for them." (Act 1).

At last, we cannot but admit that at present we are passing through a period of gloomy decadence at least insofar as our environment awareness is concerned and that this process of decay, this cankerous growth of callousness, must be stopped and reversed before it is too late.

REFERENCES

1. *vide*, for example, *Bṛhadāraṇyaka Upaniṣad*. 6th Brahmana.
2. *e.g. Śiva-Samhita*. II.
3. Expressing similar thoughts, Rabindranath Tagore writes, "It is more likely that they worshipped power not merely bacause of its utility, but because they, in their way, recognized it as truth with which their own power had its communication and in which it found its fulfillment. "*The Religion of Man* (Unwin Books, London, 1963), p.39.
4. S. Radhakrishnan, *The Principal Upaniṣads* (O. U. P., 1992), p. 30.
5. The hymns are quoted from *The Holy Vedas*, compiled and translated by Pandit Satyakam Vidyalankar, Clarion Books, Delhi, 1983.
6. *vide Puruṣa Sūktam, Rg Veda*, 10.90.1-16 and *Br. Up*. First *Brāhmaṇa*.
7. *vide Bhagavad Gītā*. 3.10-16
8. *vide Bh. Gita*. 21-23.
9. *Bh. Gita*. 3.13.
10. *Manusmriti*. 3.68-70
11. *e.g., Chāndogya Upaniṣad*. 1.2.7-11; *Br. Up*. Third *Brāhmaṇa*.
12. *op. cit*. 1.49.
13. *vide Prakriti in Sankhya-Yaga*, Knut a. Jacobson, Motilal Banarasidass Pvt. Ltd,. Delhi, 2002.

14. *vide chairanjiva Vanaushadhi* (Bengali), Kaviraj Shibkali Bhattacharya, Anand Publishers Pvt. Ltd., Kolkata, 1981.

15. *Br. Up.* 6.3.6.

16. *vide Rg Veda.*, 1.116.12; *Satapatha Brāhmaṇa*, 14.1.1.4.

17. *Br. Up.* 2.5.1-14

18 For example, *vide Nātya Śāstra* by *Bharata Muni* and *Sangīta-Ratnakāra* by *Sārngadeva.*

19. Henry W. Wells, "Sakuntala's Wateriang Pot," compiled in A Choice of Twentieth Century English Prose, ed. H.H. Anniah Gowda (O.U.P., Madras, 1978), 163.

20. Ibid p. 163-4.

21. Ibid. p. 165.

3

Environmental Protection and Awareness in Ancient India

Dr. Sanjeev Kumar*

If there is to be an ecologically sound society it will have to come from the grassroots up, not from the top down.

Paul Hawken[1]

The word " environment" comes from the French word *environ* which means surroundings, circumstances, influences. Environment means the surroundings or conditions in which a person, animal or plant lives or operates. It also denotes the natural world. The Illustrated Encyclopaedia of Modern Science defines environment as the biological term for the sum total of all the conditions, agencies and influences, which affect the development, growth, life and death of an organism, species or race. According to modern scientific notion, the physical, chemical and biotic condition surrounding an organism is its environment. Thus, the nature, which surrounds us will include the air, sky, land, soil, sea, river, forest, trees, flora-fauna, etc. Hence, I would like to quote the views of the famous philosopher, J. Krishnamurti: "What is nature ? There is a great deal of talk and endeavour to protect the nature, the animals, the birds whales and dolphins, to clean the polluted rivers, the lakes, the green fields and so on, Nature is not put together by

* **Senior Lecturer, Department. of History, Vasant College, Rajghat, Varanasi.**

thought as religion is, as belief is. Nature is the tiger, the extraordinary animal with its energy, its great sense of power. Nature is the solitary tree in the field, the meadows and the groves. It is that squirrel shyly hiding behind a bough. Nature is the ant and the bee and all the living things on the earth."[2] The concept of environment is global one and it encompasses the whole earth. Environment is not only flora and fauna but it includes the human species. It is also the store house of natural resources and one should remember that except sunlight, the supply of all other resources is fixed for all times. Resources are both renewal (as water, trees, etc.) and non-renewal (minerals, etc.) Hence, environmental problems are of two types—one relates to resource limitation and the other relates to resource misuse or misappropriation. So, there is a need for a balanced use of resources as well as conservation of natural resources.

Generally, environmental thought and awareness was seen to be the gift of western scholars. Unit very recently, studies of the environmental issues have been biased towards the western world. "There was common assumption amongst historians and sociologists that concern over such issues as conservation or biodiversity were the exclusive preserve of the affluent westerner."[3] The western scholars may have thought so, but the real situation of the Orient is totally different. The ancient base of environmental thought can be traced back as far as the pre-Vedic and Vedic times in India. The earliest track of man-environment relationship and a conscious attempt at revering nature can be seen in the proto-historic and historical periods (i.e. Indus Valley civilisation and Vedic age onwards) of India. Our ancient culture and civilisation had a strong leaning towards preservation and protection of the environment. The ancient scriptures refer to the geographical aspects, flora and fauna of various regions and give a detailed explanation of nature and environment. The ancient Indian thought depicted in the Vedic literature, Puranas, epics, etc. highlight a healthy human-environment relationship, which helped in maintaining an ecological balance. Environmental care and ethics can be seen in these ancient Indian texts. The writings of Kalidasa, Sudraka and many other scholars of ancient India give a vivid description of nature and deep man-nature relationship.

In the recent past, there has been an intense focus on environment, ecology, sustainable development and so on. The rapid scientific and technological development of mankind also ushered in the degradation of environment and ecology which is evident from the increasing pollution, deforestation, extinction and endangerment of many species of flora and fauna, ozone layer depletion leading to an adverse impact on green-house effect resulting in an increasing global warming, etc. Many of the environmental problems of today are due to the fact that we are totally divorced from our ancient culture and tradition. Nature has to be respected and followed in a congenial way. Willian Jordan has rightly said. "Nature is constantly seeking to show man that he is his own best friend, or his own worst enemy. Nature gives man the option..."[4]

In India, people were quite conscious about their immediate surroundings since ancient time. The basic principles of environmental conservation were followed and practiced in India since Vedic times. The ancient scriptures of India clearly mention that material nature *(prakriti)* and spirit *(purush)* form an inseparable part of the life support system. This system has several elements like air, water, land, flora and fauna, which are interconnected, interrelated and interdependent.[5] Nature *(prakriti)* is the complete manifestation of life in harmony. Ancient civilisation worshipped the key elements of nature-*Surya* (Sun), *Jala* (Water), *Vaya* (Wind) and *Vriksha* (Plants), recognising them as the eternal source of energy, vital to sustain life on *Prithvi* (the planet earth).

The hymns in praise of Indra, Varuna, Surya, Soma, etc. in the *Rig Veda* clearly depict a deep environmental thinking and care of the ancient sages of India. Likewise, the *Prithvi Sukta* of *Atharvaveda* speaks about man's close relationship with nature which points to positive attitude towards the environment. It says:

> Giryaste parwata himavantoranyam te prithrisyonmstu
>
> Babhrum krushnam rohini vishvarupam dhruva bhumi Prithivimindra guptam
>
> Ajitohato akshatamdhvasntan prithivimaham (XII. 1. 11)

O Mother Earth! Sacred are thy hills, snowy mountains, and deep forests. Be kind to us and bestow upon us happiness. May you be fertile, arable, and nourisher of all. May you continue supporting people of races and nations. May you protect us from your anger (natural disasters). And may no one exploit and subjugate your children.

This hymn highlights the importance of Mother Earth and also relates our feeling towards nature and environment.

The people of ancient India had a sense of wonder at their environment. Animals and birds as vehicles of gods and goddesses show a deep relationship between the different species of creation. Worship of animals, birds, trees and forces of nature depicts a deep bond between the human race and the plant-animal world.

If we compare the western materialistic culture and the ancient Indian spiritual way of life, we find that while the former has damaged the environment [like the USA who is not ready to implement in practice the recommendations of the Rio Summit (1992) and took the Johannesburg Summit (2002) lightly], the latter had always protected and nurtured the environment. It was a part and parcel of their social life as well as an integral part of their cultural ethos. Regarding the western attitude, Ranchor Prime has rightly said, "Western culture has effectively built a very elaborate cage in which the human spirit is now languishing, imprisoned by its own material excesses. Despite its sophistication, this civilisation has failed to see the inner meaning of life."[6]

The ancient Vedic worship of *pancha bhutas* (five elements)—Earth, Water, Fire, Air and Ether (or Space/Sky) is a tribute to nature. The *Yajurveda* clearly states: "May my five elements—Earth, Water, Fire, Air and Ether (or Space)—become purified. May I become the light (or the Divine Principle of life and intelligence), free from impurity and evil. To this end I offer this oblation into the consecrated fire."[7] These five elements complement each other to create a harmonious world and a proper use of the *pancha bhutas* by mankind can make our environment a better and healthy one as it was in the past. The present day abuse of our environment is proving to be ruinous for mankind. Many of the principles for environmental care and protection enunciated at the Rio Earth Summit in June

1992 are clearly depicted in Vedic literature, that is environmental conservation was practiced by people of ancient India.

Reverence for Earth

In Vedic literature, mother earth *(Prithvi)* is personified as a goddess. The relation between the earth and her inhabitants is seen as that of a mother and her children. Such references are also found in many inscriptions. The Parbatiya copper-plate Inscription of Vanamalavarman (c-835-60 A.D.), a ruler of Assam, refers to the earth as goddess earth.[8] Similarly, the Tanjavur Rajarajesvara Temple Inscription of Rajendra Cola I, (1012-44 A.D.) refers to the goddess earth.[9]

Nature and Cosmic Force

Lord Vishnu's body is compared to different parts of the nature. "It is said that the Oceans are Vishnu's waist, the hills and mountains are his bones, the clouds are the hairs on his head and the air is his breathing. Rivers are his veins, the trees are the hairs on his body, the sun and moon are his two eyes and the passage of day and night is the moving of his eyelids."[10]

The following quotes from the *Bhagvadgita* illustrate the fact that Lord Krishna is relating himself to various forces of nature like the fire, wind, sun, moon, rivers, mountains or the various living organisms like the tree, animals, birds and fish which altogether form the bio-diversity of the planet earth. Lord Krishna in the *Bhagavadgita* says: "I am... the radiant sun *(ravi)* among the luminaries; I am the glow of the Maruts (the forty-nine wind-gods) and the moon *(sasi)* among the stars." (X. 21) "... Among the eight Vasus, I am the god of fire *(pavaka)*; and among the mountains, I am the Meru (X. 23). ".. among the immovable, the Himalaya" (X. 25) "... Among all trees, I am Asvattha (the holy fig tree) ..." (X. 26). "Among horses, know me to be the celestial horse Uccaihsrava .. among mighty elephants, Airavata..." (X. 27). "...among cows, I am the celestial cow Kamadhenu...among serpents I am Vasuki" (X. 28) "...so among quadrupeds, I am the lion, and among birds, Garuda," (X. 30). "Among purifiers. I am the wind *(pavan)*; Among fishes, I am the shark, and among streams, I am the Ganges (Jahanavi)." (X. 31).

I

Forest and Trees

First, let us discern our ancient tradition regarding forest and trees—its importance and the way people conserved it. In the 21st century, we find that there is an acute pressure on forest and wildlife. As the human population and its domination over the Earth have increased, the threat to forests and naturally evolved life forms has also increased. As the human population continues to rise, more and more land is brought under direct human control for agriculture (and other purposes like habitation, etc.), the amount of natural vegetation diminishes and with it the space available for the species which live in such habitats. The vast expanses of tropical forests have become increasingly threatened in the last few decades.[11]

In the *Varnasrama* system there were four stages of life—*brahmacarya, grihastha, vanaprastha* and *sanyasa*. Except *grhastha* the other three stages were closely linked to forest. In the first stage, a person had to be enrolled in *gurukula* which was situated in the forest area. The *rishis* lived in forests with their disciples imparting Vedic knowledge in a natural setting. The students learnt to live in forest with harmony. A congenial man-nature relationship was built up. The Vedic teachers attached great importance to trees beneath which they delivered their daily discourse to the students. In the third stage of life, that is *vanaprastha,* a person after fulfilling his duties as a householder, left his home for the forest to lead a retired and recluse life. In this stage, he was required to do penance and meditation to free his soul from material things. In the last stage, he became a wanderer *(sanyasin)* renouncing all his worldly ties. In these three stages, meditation was a part of daily life and meditation took place mostly in forest where people found so much of serenity, peace and inspiration. In contrast to the ancient society, the modern society has left the forest far behind. In fact they are destroying it and the rate of deforestation is alarming. The required forest cover of 33% is still far behind. The Forest Survey of India report 2001 states: "The forest cover of the country, as per the present assessment, is 63.73 million hectare constituting 19.39 per cent of its geographical area."[12] The latest "State of Forest Report 2001" released on 22 April 2003 says that the forest cover was assessed at

675, 538 square km., 20.55 per cent of the country's geographical area.[13] But according to National Remote Sensing Agency, the forest cover in India is not more than 11 per cent.[14]

In the ancient period, India was much better afforested. Hiuen-tsang who visited India in the 7th century A.D. mentioned in his account that dense forest lined the banks of the river *Ganga* for many miles on end. The countryside was much wilder as compared to today.[15] The planting of trees like Mango, Neem or Banyan on the roadside is evident since ancient times in India. These trees not only provided shade or relief from scorching sun but also played an important role in Earth's greenhouse effect. Afforestation and digging of wells have been considered to be two great acts of merit *(punya)* since ancient times. Asoka's policy of *Dhamma* included such humanitarian activities. Asoka's 7th Pillar Edict says, "Thus speaks the Beloved of the gods, the king Piyadassi: On the roads I have had banyan trees planted, which will give shade to beasts and men. I have had mango groves planted and I have had wells dug..."[16]

The importance of forest was recognised during the Mauryan period. The *Arthasastra* of Kautilya mentions a Superintendent who looked after the forest and the forest produce. The clearance of forests was regulated and *Arthasastra* suggests deployment of soldiers or guards to check unnecessary damage to forest. The unnecessary damage to forest was prohibited. Asoka's 5th Pillar Edict specifies: "Forests must not be burned in order to kill living things or without any good reason."[17] Even in the Vedic period, if forest was cleared for cultivation or inhabitation, the tradition required that another kind of forest be established to make up for the deforestation. According to tradition it was not trees that should be in the village but vice versa i.e. the village should be among trees. The *rishis* of ancient India lived in the forest where they built their *asramas* or hermitages. The presence of these saints guaranteed the protection of the forest and the wildlife. No animal or plant was harmed. The dense forest was conserved and its natural setting was never disturbed. Human activities were limited to the collection of dry wood, leaves, forest produce and a little amount of green timber. The village communities residing near the forest conserved it generation after generation. It became a part and parcel of their life.

Medicinal Herbs

The sages in ancient period also studied and recorded the herbal and medicinal properties of the forest. Some trees gained special significance and hymns, prayers, etc. were composed about them. Medicinal plants and herbs were widely used for protection of health and were planted and protected as a part of ecosystem. *Ayurveda* (sanskrit ayur = life, veda = Knowledge or science) is the ancient system of health management whose medicines are mostly based on plant extracts. A medicinal plant can be defined as "any plant which in one or more of its organs contains substances that can be used for therapeutic purposes or which are precursors for chemo pharmaceutical semi-synthesis."[18] Medicinal plants are the most ancient source of medicine and the plant-based medicaments are still used today for treating varieties of diseases. India and China are considered to be the treasure houses of traditional medicines as thousands of species of plant growing in these countries have contributed to its development. This was possible due to the availability of a wide variety of geographical and ecological conditions. The development of traditional herbal medicine system can be traced in the ancient text like *Caraka samhita* and *Susruta samhita*. The *Caraka Samhita* mentions 340 herbal medicines. Many of the current medicinal plants in use can be traced back to the Vedic age. The Vedic literature contains a large number of terms which denote plants and their parts. Many hymns in the *Vedas* were composed involving the medicinal plants; the *Osadhi Sukta* of the *Rig Veda* (X. 97.1-23) and the relevant portion of the *Atharvaveda* (*Saunak* School, VIII. 7.1-8, XI. 6.16-17) testify to this fact. The Vedic people acquired the knowledge of plants properly through the close association of animals and many medicinal plants bear their nomenclature on the basis of animals and their parts.[19] A few examples are:

	Medicinal Plants	*Name of Animal*
1.	Ajaji (Jeera)	Aja (Goat)
2.	Avipriya (Shyama)	Avi (Sheep)
3.	Asvagandha	Asva (Horse)
4.	Musakarni	Musaka (Rat)

The conservation and propagation of the medicinal herbs and plants was a part of environmental ethics in ancient India. The 2nd Major Rock Edict of King Asoka clearly states. "....Medicinal herbs whether useful to man or to beast, have been brought and planted whenever they did not grow, similarly, roots and fruits have been brought and planted wherever they did not grow..."[20] The Paschimbhag Copper-plate Inscription of Srichandra (c. 925-75 A.D.) refers to a particular forest Lalambi-vana which was famous for its medicinal herbs.[21] Likewise, the ninth mandala of the *Rig Veda* is exclusively devoted to Soma, the lord of forest *(vanadevta)* and medicines *(osadhipati)*. The people of ancient India recorded the flora and fauna vividly. The Khuda Baksh Oriental Public Library, Patna, in its rich collection has a richly illustrated botanical guide dating back to the sixth century A.D.[22]

Veneration of Trees

We find that trees were worshipped or venerated in India since the Harappan age. Trees were worshipped in material form and also endowed with human shapes and attributes. A terracotta seal bearing the sketch of *Pipal* tree *(Ficus religiosa)* has been found in the excavation. Perhaps, the *Pipal* was regarded as sacred which continues to be worshipped even today. Among other trees, the *Acacia* is represented enclosed in a platform or with a railing round its base, on a number of seals from Harappa. The Vedas also mention a number of trees which were considered sacred. *Asvattha (Pipal)* is mentioned in the *Rig Veda* (I.164.20) and is designated as the king among the *Vanaspatis* (medicinal plants) in *Aitareya Brahmana* (VII. 32, VIII. 16). This tree was the chief 'sacred tree' of the Buddhists and was venerated by them as Lord Buddha received illumination *(bodhi)* under the shade of a *Pipal* tree at Bodh Gaya. Thus, *Pipal* became the *'Bodhi Vriksha'* for the Buddhists and it was a sin to cut or harm any Pipal tree. In *Bhagavad Gita,* Sri Krishna identifies himself with the Pipal tree. He says, "Among all trees, I am the *Asvattha* (the holy Pipal tree)."[23] The planting of trees *Pipal, Neem,* Banyan, Mango, *Asoka, Tulasi,* etc. was always considered a pious act and many of them were associated with religious acts. The *Veraha Purana* says: "One who plants one *pipal,* one *neem,* ten flowering plants or creepers, two pomegranates, two oranges and five mango trees will not go to hell" (172.39). Trees were considered so important in ancient India that many of the law books including *Manusmriti* described illegal cutting of trees as a punishable crime.

Manu-Smriti mentions a host of flora and fauna. The classification of trees and plants is seen in this text (I. 46-48). Manu further says that the trees and plants have consciousness within and can feel pleasure or pain (I.49). The same notion is mentioned in the *Rig-Veda* too. (X.97-21)[24]. The mention of consciousness may be an indirect way of securing a compassionate and balanced approach towards the plants and trees. Manu also prescribes different punishments for causing injury to plants.

Man-Forest Relationship

The man-forest relationship can be viewed as :

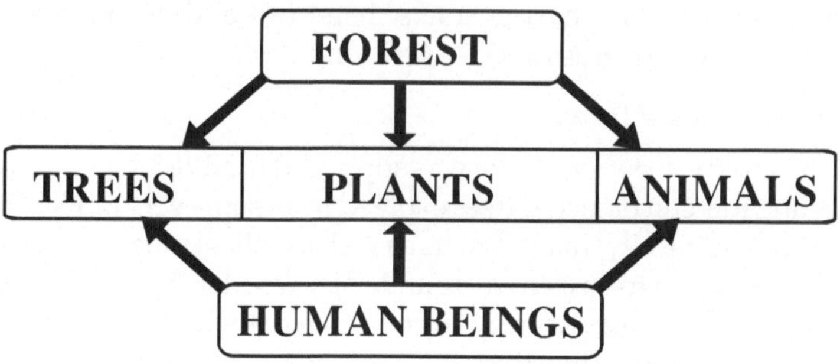

The ecological relationship built between human society and the forest community comprising trees, plants and animals is based on sustainable co-existence which is free from exploitation. This subtle thinking evolved due to the Indian tradition of respecting and valuing trees. The earth not only belongs to mankind but also to other species living on earth. The Vedic scriptures teach us to protect trees, animals and all species which live on earth. "Forest, at one level, means the world. It includes the whole creation. You are also part of the forest. It is not that you are outside the forest. You can reorder it, but you cannot bypass it."[25] The importance of forest is vividly expressed in the statement quoted above. Forests are our "green lungs" as well as our "green medicinal store." The glory of trees is beautifully stated in this verse from the Varaha Purana: "Trees have five sorts of kindness which are their daily sacrifice. To families they give fuel, to passersby they give shade and a resting place ; to birds they give shelter ; with their leaves, roots and bark they give medicines. (162.41-42).

The Gupta and post-Gupta period witnessed the granting of land to the brahmanas, temples and officials. The granting of land in the forest region to the brahmanas was an attempt to build a man-forest relationship through a judicious use of forest produce. The Tipperah Copper-plate Inscription of Lokanatha (664A.D.)[26] refers to the grant of a piece of land in the forest region to a brahmana for the maintenance of the god Anantanarayana. This grant was made in the region of Bengal *(Gauda)*. Likewise, the Madhainagar Copper-plate Inscription of Lakshmanasena (c. 1179-1206 A.D.) of Bengal mentions the grant of a village as a gift. It further states "the gift was made together with jungles and bushes or branches, land and water, pits and barren tracts, betel nut and coconut trees, preserves *(yuti)* of grass and pasture land..."[27]

Sacred Groves/Trees

Each city in ancient India was supposed to have its own sacred trees and this cult of sacred trees evolved from some very old fertility cult. The capital city, many towns as well as villages also possessed one.[28] The sacred trees were surrounded by wooden or stone barriers to protect them against any attack. Since ancient times, many traditional societies (like the tribal societies, etc.) have revered and worshipped nature and considered certain plants as sacred. "Sacred groves" dedicated to a deity or god was well protected and worshipped. This ancient tradition is still witnessed in states like Kerala, Karnataka, Tamilnadu, Maharashtra, M.P., Rajasthan, Orissa, Himachal Pradesh and Jharkhand. These sacred groves are rich pockets of greenery. In spite of deforestation process continuing unabated, some of the sacred groves are still intact, conserving rich biological diversity. "The maintenance of sacred groves is and example of a traditional practice that has contributed to the conservation of nature, though in a small measure."[29]

There are numerous references to gardens and sacred groves in ancient Sanskrit, Prakrit and Tamil literature. The Tamil epics, *Silapadhikaaraam* and *Manimekalai* refer to many parks and gardens. There where the *Nandanvanas* (Nandavana), *Devavanas* or sacred groves *(Kovil Kaadu* in Tamil, *Kaavu* in Malayalam, *Devarakaddu* in Kannada and *Deoran-devavana* or *Oran* in Maharashtra and Rajasthan). The gardens were closely associated with religion. Nanditha Krishna writes, "The *Devavana* or *Kovil Kaadu* (sacred groves) was dedicated to the mother goddess. The sacred groves were repositories of local plant species and mini biosphere reserves: instead of offering a flower or a fruit, all the trees in the grove were dedicated to her. In the grove there was a pond, in which grew medicinal plants. Cutting trees, even a branch or a leaf was taboo. *Nadavanas,* on the other hand, generally contained a choice of flowering plants essential for *pooja*. What is wonderful in this matching of each god to a plant or flower is the strong ecological message: conserve biodiversity by preserving a variety of trees or plants."[30]

The oldest Tamil literature referred to as the Sangam literature recognised five ecological regions *(tinai) - kurinji* (hills and mountains), *mullai* (woodland or pasture lands), *marudam* (cultivated plains/agricultural lands), *neydal* (coastal areas) and *palai* (desert). Each region had a separate ecosystem having its own appropriate flowers, animals, people and even gods. "While the natural local vegetation characterised the last three, special efforts were made by local rulers to plant a variety of flowering plants in the *kurinji* and *mullai*. Kings and noblemen donated land, money, gold and watering facilities (tanks or wells) for the establishment and maintenance of *Nandavanas.*"[31]

Gardens

The provision of garden in and around temples is attested by inscriptions. The Bhubaneswar Stone Inscription of Bhavadeva, minister of Harivarman (c. 1090-1110A.D.) clearly mentions that a tank was excavated in the front of and a garden lay outside the temple."[32]

Gardening, horticulture, etc. played an important role in ecology and environmental balance. In ancient India, the family's private life centered on the garden and it's out buildings. In garden, they grew vegetables as well as the medicinal plants needed for treating common ailments. "The Hindu epic tradition concentrates on forests but refers to *vatikas* (gardens) as well. These are the locales where protagonists of epic tales relate to nature and through it to God."[33] The pleasure grounds surrounding Indraprastha are described in the *Mahabharta.* The Buddist text *Lalitavistara* mentions 500 gardens around Kapilvastu. The stories narrated in the *Mahabharata* and *Ramayana* are closely linked to the forest. Seema Alvi writes "In the *Ramayana,* forests form the context in which not only do the protagonists relate to nature, but they also negotiate the spiritual world as well. The *Ramayana* has vivid descriptions of Rama's *vanavas* (exile in forest). The forest is described as being beautiful and idyllic, full of trees of fruit and nuts, birds and beasts. Rama's relationship to his environment is exemplified in his numerous hunting forays and his search for fruits and herbs for survival."[34] In Ramayana, the forest is also the residing place of *rishis* and their *asramas* (hermitages) are described as places of peace and silence. The sages carried out the *yajnas* and *havans* which had a purifying effect on the environment. The forest in fact is the meeting point of the sacred and the profane where man and nature come in close ecological interaction. The epics also have the concept of gardens—the *Vatikas. Asoka vatika* as described in the Ramayana is a place full of plants and trees which are ecologically balanced with birds, animals and man. "The imagery of the forest and the garden also reflect the two varied ecological contexts in which man continues to relate with nature even as his society moves from one social order to another."[35]

Gardens were an indispensable feature in house and town planning in ancient times. The Aryans were lovers of nature and they evolved a refined art of gardening which continued in later periods too. In Vatsayana's *Kamsutra, "Vrkshaymveda"* is mentioned as one of 64 arts. It included the construction and maintenance of gardens and parks for health, recreation and enjoyment. In Jain canonical texts too, among the important parts of a city mentioned are pleasure gardens *(arama),* gardens *(ujjana)* and parks *(vapi).* Garden was seen as a source of joy and happiness throughout the ancient period. Ancient rulers of India built pleasure gardens for themselves. The ancient Indian greatly loved flowers and trees. Megasthenes describes with wonder the beautiful parks surrounding the palace of Chandragupta Maurya and many references in Sanskrit literature show that wealthy citizens had gardens attached to their houses and often larger parks in the suburbs containing pavilions in which they spent much of their leisure.[36]

The early Buddhist period saw the transition from royal to public gardens *(nagaruavana)* at many places. There are references to private gardens attached to houses. Vatsyayana in *Kamasutra* state "attached to every house there should be a *vrksavatika* or *puspavatika*, a garden...in the middle."[37] With gardens and parks emerging as an important backdrop to the social life in ancient India, horticulture *(Udyanavyapara)* developed as a discipline.

We have inscriptional evidence referring to donations made for the maintenance of the gardens. A royal edict specifically mentioned that the water of the *Porporavai* tank belonging to the temple of *Natrunainatha* was forbidden for any use except for watering the *Nandavanam*. An inscription of Kulothunga III refers to a committee constituted for the management of the *Nandavanam* of the Nataraja temple at Chidambaram.[38] People's concern for the protection of these gardens is depicted both in literature and inscriptions. There were temple festivals which laid stress on the sanctity of the plants and its conservation.

Thus, religion played an important role in conservation and protection of the environment in ancient period. Forests, gardens, parks became a part and parcel of perfectly balanced greenery and ecological balance. They city in ancient India had many "green zones" as a large number of canals, fountains, public bathing pools, ponds, and natural or artificial lakes were located in its area. Pleasure gardens, parks alternated with alleys bordered by trees and mango groves and it was a familiar sight in ancient period.[39]

II

Wildlife and Animals

In this section, we will deal with the wildlife and animals in general. We have to accept the fact that there exists a wide gap between the position occupied by animals in our cultural and religious tradition and the present state of Indian wildlife. Almost all deities were associated with one or the other animal and the animals were revered almost at par with the deities. The ancient

Indian texts are full of references towards the conservation and protection of animals and birds. It was made more fruitful by according them a place in the religious sphere. The cow, bull, rat, owl, snake, tortoise, swan, lion, elephant, etc. were accorded an exalted position as the vehicles *(vahana)* of various deities. The society, thus, respected them and also preserved them with religious fervor and sanctity. The carvings on the gold coins of king Sasanka (c. 600-625 A.D.)[40], ruler of Bengal, depict the flora-fauna-deity relationship. Lord Siva is shown reclining on couchant bull on one side of the coin. The reverse side shows goddess Lakshmi seated on lotus, facing, with an elephant on either side sprinkling water on her; lotus in her left hand resting on the knee. In India, there is an ancient bonding with wildlife and animals. Our folk culture, art and sculpture depict the deep sense of admiration for the animal world. They are widely depicted in Indian art. The major centers of art during the 2nd century B.C.—Sanchi, Bharhut and Bodhgaya represent animals in abundance. But the present day reality is far from being satisfactory as many species of animals have become extinct or are on the verge of extinction.

In ancient times, tigers roamed throughout India. Likewise, the lion was common in western India. But in present times, they have become endangered species. The elephants which roamed the length and breadth of India are now in need of protection. The cheetah is now extinct in India. In the present times, there in an urgent need to protect the wildlife and various species of animals. Animals are an important pillar of agriculture and other economic activities. The *Athrvaveda* (अर्थर्विद संहिता भाग-२, औटुम्बरमणि सूक्त-१) clearly advocates the need to increase the animal wealth. It further states the importance of the cow as giver of milk as well as cow dung which can be used as a fertilizer (अर्थर्ववेद संहिता भाग-२, औटुम्बरमणि सूक्त-३). If we view these two hymns of *Atharvaveda* in the present context, then we realize the wisdom of our ancient seers. The chemical fertilizers and pesticides have added to the environmental degradation of the soil as well as the vegetation. So, the focus has once again shifted to bio-fertilisers and eco-friendly alternatives. The cow-dung has once again appeared as a solution to many of the problems of agricultural productivity visa-vis the environmental protection.

India harbours over 50 per cent of the wild elephant and over 20 per cent of the captive elephants in Asia. It is closely associated with the religions, myths, history and cultural heritage of our country. It is considered and auspicious symbol in Hinduism, Buddhism and Jainism. It is symbol of prosperity and abundance. Elephant symbolically represents the birth of Buddha. The Jaina Tirthankaras were also linked to animals and among them Ajitanath was associated with elephant. In Hinduism, there is prevalence of the cult of Gaja-Laksmi. The seals of the Sarabhapuriya rulers of Daksina Kosala (who ruled over the present day Raipur, Bilaspur, Sambalpur in the post-Gupta period) show a standing figure of Laksmi on a water-lily flanked on both sides by an elephant with its trunk lifted up to pour water over Laksmi's head.[41] The representation of Gaja-Laksmi shows a close association of elephants with goddess Laksmi.

Elephant is well represented in the Indian tradition. It is also an indicator of the health of the environment. A good population of elephants and other wild animals means a good forest cover which in turn means rains and a perfect ecological balance for sustenance of life.

The concept of incarnation *(Avatar)* in the Hindu religion is also linked to animals. The incarnations of *matsya* (fish), *kurma* (tortoise), *varaha* (boar), *narsimha* (man-lion) were all associated with the animal world. This shows that animals have an important role on this earth. Even Lord Rama and Krishna were closely associated with monkeys and cows respectively. Lord Hanuman was the Moneys god who is still worshipped and revered by Hindus.

A compassionate attitude towards animals is a part of our ancient tradition. The *Vishnupurana* says: "God Kesava is pleased with a person who does not harm or destroy other non-speaking creatures or animals." [3-8-15]. The Asokan edicts, too, advocated a compassionate attitude towards animals and his policy of *Dhamma* explicitly stated this notion. The 1st Major Rock Edict says, "...no living thing having been killed is to be sacrificed...Formerly in the kitchens of ...king Piyadassi, many hundreds of thousands of living animals were killed daily for meat. But now...only three animals are killed, two peacocks and a deer, and the deer not invariably.

Even these three animals will not be killed in future."[42] He perhaps opened the first hospital for animals. He also gave up hunting and refrained from eating living beings. In the 5th Pillar Edict, Asoka orders that certain animals are not to be killed on certain days, and others are not to be killed at all. The *Arthasastra,* too, says, "Cattle such as a calf, a bull or a milk cow shall not be slaughtered." (II.26). Manu also refers to various types of animals and specifically mentions the animals whose meat is permitted. Most of the birds mentioned in the *Manu-smriti* are forbidden as meat and their killing is considered to be sinful. (V.12, 14; XI. 131, 135 etc.)

Lord Buddha is known for his compassion towards animals and all living beings. Buddhism tried to promote the welfare of people as well as animals. The notion of *Ahimsa* (non-violence) is deeply rooted in the Indian tradition. The spiritual tradition points to the oneness and sacredness of all living beings. "One should treat animals such as deer, camels, asses, monkeys, snakes, birds and flies exactly like one's own children. How little difference there actually is between children and these innocent animals.[43]

III

Water Resources

Rivers

Rivers have played an important role in human civilization. The ancient Egyptian, Chinese, Mesopotamian civilization arose in a particular geographical setting benefiting from the river banks. Our own Indus valley civilization mainly blossomed on the banks of the river Indus and its tributaries. Further, the rise of 16 *Mahajanapadas* in the sixth century B.C. was mostly confined to the Ganga-Yamuna doab region. Thus, the rivers have been the cradle of civilization in many parts of the world. Rivers were considered holy and worshipped in ancient India and this practice continues even today. Ganga is the cultural and ethical signpost of our ancient culture. But the situation has totally changed. The rivers need conservation and protection as they have been victims of unabated pollution. So, there is a need to learn from our ancient tradition and check the increasing pollution of our rivers as government programmes themselves are not competent to solve the problem.

The river has occupied a central place in Indian cultural arena. The major rivers of India have been projected in two distinct groups: Indus (Sindhu), Brahmaputra, Son (Sonabhadra), Ghaghra, etc. which are broad but not too deep are treated as male while Ganga, Yamuna, Saraswati, Narmada, Kaveri and Godavari which are long and deep are treated as females.[44] The river has been revered and worshipped and treated as a sacred figure since time immemorial. The *Nadistuti* or hymn or rivers in *Rig Veda* is a testimony to this fact. "The *nadi-devatas* or river-goddesses of later times evolved from the Vedic *apsaras:* more benign that their precursors, theirs is an iconography of fertility and abundance. Four images of *nadi-devatas,* each accompanied by a special totemic vehicle, appear in the temple of the sixty-four *yoginis* at Bhera Ghat, on the *Narmada:* here, we see Ganga with her *makara* or horned crocodile, Yamuna with her tortoise, Saraswati with her peacock, and Narmada herself, on a *makara* pedestal. Each river goddess carries the archetypal vessel of plenitude, the *purna-ghata.*"[45] The reverence accorded to these rivers and their association with animals as their vehicle or *vahana* is amply testified here. So, the fact that the rivers which were sacred and worshipped became defiled and polluted with the passage of time is a disturbing phenomenon. The ancient texts lay emphasis on maintaining the purity of the rivers and even prescribe punishment for polluting or making the river dirty. It included penance, outcasting, fine, etc. The *Padma Purana* clearly states that spitting, urinating or defecating, dumping filth or waste in the water of Ganga is a great sin. Whosoever does it will incur the sin of *Brahmahatya* (killing of a brahamana) and will go to hell.[46]

War Over Water

The Indian tradition and mythology is full of episodes narrating war over water which was fought over location of dams or diversion of rivers. Indra, the god of rain killed demons like Vritra and Namuchi who were obstructing the free flow of water. The Battle of Ten kings, mentioned in the *Rig Veda* was over the sharing of river water. "The struggle of those attempting to preserve local habitats shaped by long intimacy with riverine environments, against those who would destroy them in the name of totalizing economic and political processes, was being waged even four millennia ago in the subcontinent."[47]

Water Harvesting and Conservation: Tanks and Reservoir

Water tank or reservoir of rainwater of varying size was an essential part of the traditional ecology of the Indian villages. It was a wonderful way of water harvesting and conservation. In ancient India, these tanks were constructed near temples or sacred pilgrimage place by kings, royal families and wealthy benefactors. The construction of dam for irrigation was another way of water management. The Janagarh Rock Inscription of Rudradaman dated A. D. 150 mentions that a dam was constructed on the Sudarsana lake in the neighborhood of Girnar during the reign of Chandragupta Maurya. Conduits were constructed during Asoka's reign. The lake continued to function till the Gupta age as an inscription of Skandgupta informs us that it was repaired during his reign.

Construction of tanks was considered to be a meritorious act *(punya)* in the ancient period. The epics and the Puranas amply testify to this fact. Mahabharata in its *Banadharma-parva* says *Durlabham Salilam Tata Visheshena Paratra Vai* meaning water is scarce, hard to get, rare, particularly on other planets.[48] Hence, it pleads for water conservation through rain harvesting and water reservoirs or tanks. The epic says that one who "conserves water earns religious merit *(punya)* equal to performing *Agnihotra* ritual. It further states that the tank where rain-harvested water lasts upto *Sharad ritu* or autumn will earn religious merit for its digger equal to donating one thousand cows. The tank in which water conserved in the rainy season lasts up to *Hemanta ritu* or early winter earn for its creator religious merit equal to giving generous gift or donation of gold to Brahmin in a *yajna.* The tank in which water remains intact till *Shishir ritu* or winter earn for its digger the *punya* or religious merit equal to performing *Agnisthama yajna.* The tank in which rain-harvested water lasts upto *Vasant ritu* or spring and serves the thirsty earns for its maker the *punya* equal to performing *Atiratra yajna.* And finally, the tank in which stored rain water lasts upto *Grishma ritu* or summer season earn for its digger the religious merit equal to performing the great *Ashvamedha yajna."*[49] There are inscriptional evidences for the construction of tanks. Tanks were a means of conserving water. The Apshad Stone Inscription of Adityasena (c. 660-75 A.D.) refers to the construction of a tank in the region of Gaya in Bihar. Such references are found in many of the inscriptions in the region of Bengal, Bihar, Orissa, Rajasthan, etc.[50]

The India tradition emphasizes the purity of water and says that it should never be contaminated with human waste. Water is considered to be the symbol of purity. "One should not cause urine, stool or mucus to enter water. Anything mixed with these unholy substances, or with blood or poison, should never be thrown into water."[51]

IV

Sustainable Tradition in India

The Indian tradition has always believed in a sustainable environmental development. Planting of trees, herbs, etc. was very important in ancient times and people believed in afforestation as well as kitchen gardening, horticulture and gardening. They lived and followed nature. The Indian spiritual tradition of Hinduism, Buddhism and Jainism treat nature as the mother and goddess and "nature does not need to be controlled or transformed or decorated, she needs to be accepted in her entirety as a whole ... She can be loved and celebrated as she is ... But she is alive and must mot be diminished or reduced."[52]

Buddhism as a religion and a way of living throws light on the environmental crisis and the ways and means to tackle it for a peaceful co-existence of human beings. The entire Vedic literature points to a *harmonious man-nature relationship*. The Vedic sages composed hymns is honour of various forces of nature like the sun (Surya), wind (Vayu), water (Varuna), Fire (Agni), rain (Indra), storm (Maruts), etc. The reference to various gods and goddesses in Vedic literature where they represent one aspect of nature points to the fact that different facets of nature have been given importance since ancient times. It shows a deep feeling for nature.

Environmental concern and preservation was a part of our folk and rural culture. India, which has been predominantly an agricultural country, was closely attached to the environment. Many of the festivals observe in the past and even today are linked to the environment. Deepawali, Bihu, Vaisakhi, Onam, Bat-Savitri are some of them.[53] The folk culture of India also talks of various pollution of soil, water, air, etc. and the ways to preserve and conserve the same by a balanced and sustainable approach.[54]

Forests hold the key to India's water, ecological and food security. The loss of forest means flood, drought and soil infertility leading to collapse of life support system. Forests are the largest repository of terrestrial biodiversity. Forests have provided sustenance to rural and tribal communities who had a sense of reverence towards the forest, trees and nature. The Jarawas (a bunch of 250 odd tribals living in the south and middle Andaman Islands) still "maintain a lifestyle in total harmony with their environ and prefer their forest life. They only take as much as they need for the day from their forests and never disturb nature's regeneration process."[55] Their life teachs us what Gandhiji repeatedly said, "The earth provides enough to satisfy everyone's need, but not for anyone's greed."

The archaeological evidence suggests that great civilizations such as the Sumerians and the Mayans, met devastation at least in part by failing to live in harmony with the natural environment. According to scholars like Fairservis, the Harappan civilization declined due to ecological imbalance. This imbalance was caused because the human and cattle population depleted the scarce resources, the forests and food. The disappearance of forest led to more floods and droughts. The decline of the ancient civilizations should act as an eye-opener to all of us so that we learn from history and history in not repeated in terms of collapse of human civilization. If we don't learn from history, then history is bound to repeat itself and it will be a doomsday for mankind. We should remember the words of Jose Ortega Gasset, "I am I plus my surroundings, and if I do not preserve the latter, I do not preserve myself."[56] So, there is an urgent need to look back towards our ancient tradition, which stood for a balanced co-existence of mankind with nature. The civil society must play a greater role in afforestation, tree-planting and other environment protection and sustainable development activities.

NOTES

1. Sacred Space, *The Times of India,* New Delhi, 8th April, 2002.
2. *Wake up India,* Vol. XXIII, No. 1, Jan-March 2003, p. 10.
3. Ramachandra Guha & J. Martinez-Alier, *Varieties of Environmentalism: Essays North and South,* OUP, Delhi, 1998, Preface.

4. *Wake up India,* Jan-March 2002, p. 16.
5. *Cultural Heritage of India,* Vol. I, II, III, Ramakrishna Mission, Calcutta, 1972 in R. B. Jain, *Public Administration in India-21st Century Challenges for Good Governance,* New Delhi, 2002, p. 367.
6. Ranchor Prime, *Hinduism and Ecology,* Motital Banarsidass, New Delhi, 1996 (reprint), p. 4.
7. *Taittiriya Aranyaka of Yajurveda,* 10-66-1.
8. D. C. Sircar, *Select Inscriptions,* vol, II, Motilal Banarsidass, Delhi, 1983, p. 18.
9. Ibid., p. 633.
10. Srimad Bhagavatam, 2.1.32-33, quoted in R. Prime, *op.cit.,* p. 26.
11. J. L. Chapman & M. J. Reiss, *Ecology Principles and Application,* Cambridge Low Price Edition, 1995, p. 257.
12. *Hindustan Times,* New Delhi, March 23, 2001, p. 5.
13. "Is Our Forest Cover up ?", *The Times of India,* New Delhi, April 24, 2003.
14. *Paryavaran Chetna,* September 2001, p. 6.
15. A. L. Basham, *The Wonder That Was India,* Rupa & Co., Delhi, 1995, p. 192.
16. Romila Thapar, *Asoka and the Decline of the Mauryas,* Oxford University Press, Delhi, 1993 (reprint), p. 265.
17. Ibid., p. 264.
18. V.B. Pandey, "Evaluation of Medicinal Plants and Plant Base Medicaments", *Amrit,* Vol. I, No.1, Jan-March 1988, p. 11.
19. Dr. Jyotir Mitra, "Some Current Medicinal Plants of Vedic Antiquity", *Amrit,* Vol. I, No. 1, Jan-March 1988, pp. 55-59.
20. Romila Thapar, *op. cit., p. 251.*
21. D. C. Sircar, *op. cit., p. 92.*
22. *India Today,* 22 April, 2002, p. 8.
23. *Bhagavad Gita,* Chapter X. 26.
24. S. C. Banerji, *A Companion to Dharmasastra,* D. K. Printworld, New Delhi, 1998, p.184.
25. Banwari, Quoted in Ranchor Prime, *op. cit., p. 8.*
26. D. C. Sircar, *op. cit., p. 28-29.*
27. Ibid, pp. 124-125.
28. Jeannine Auboyer, *Daily Life in Ancient India,* (translated from French by Simon Watson Taylor), Asia Publishing House, London, 1965, pp. 121, 138.

29. S. M. Nair, "Rich Pockets of Green", *The Hindu folio,* June 4, 2000, p. 49.
30. Nanditha Krishna, "Retreat of the Gods,", *The Hindu folio,* June 4, 2000, p. 18.
31. Ibid.
32. D. C. Sircar, *op. cit., p. 105-106.*
33. Seena Alvi, *The Hindu folio, June 4, 2000, p. 12.*
34. Ibid., p. 15.
35. Ibid., p. 15.
36. A. L. Basham, *op. cit.,* p. 204.
37. Seema Alvi, *op. cit.,* pp. 16-17.
38. Nanditha Krishna, *op. cit.,* p. 19.
39. Jeannine Auboyer, *op. cit.,* pp. 124-125.
40. D. C. Sircar, *op. cit., p. 28.*
41. Sanjeev Kumar, *A Socio-Religious History of Daksina Kosala: A Study Based on Inscriptions,* M. Phil Dissertaion (unpublished), JNU, Delhi, 1994.
42. R. Thapar, *op. cit.,* p. 250.
43. Srimad Bhagavatam, 7-14-9, quoted in R. Prime, *op. cit.,* p. 51.
44. Manoj Das, "Of myths and legends", *The Hindu folio,* July 1, 2001, p. 14.
45. Ranjit Hoskote, "Looking for the river", *The Hindu folio,* July 1, 2001, p. 6.
46. Padma Purana, Kriyayoga, 8.8-10, quoted in *Paryawaran Chetna,* Jan-Feb 2003, p. 10.
47. Ranjit Hoskote, *op. cit.,* p. 8.
48. M. L. Varadpande, "Conserve water, says Mahabharata", *The Hindustan Times,* Delhi, 3 July, 2000, p. 14.
49. Ibid.
50. D. C. Sircar, *op. cit.,* pp. 44-45, 102-03, 105, 115, 116, 236.
51. Manu-Smriti, 4.56, Quoted by O. P. Dwivedi in *Environmental Crisis and Hindu Religion,* Gitanjali Publishing House, Delhi, 1987.
52. Ravi Ravindra, *Science and Spirit,* Paragon House, New York, 1991, p. 38.
53. For details on this aspect of environment, kindly refer to an article by Krishna Kalki, "पर्यावरण चेतना का प्रतीक ग्रामपर्व वट सावित्री, कुरुक्षेत्र", May 2004, pp. 23-26.
54. For details, refer to Dr. Narmada Prasad Gupta's article, "प्रदूषण की समस्या और लोकसंस्कृति", गगना चल , Oct-Dec 1999, pp. 21-30.

55. Rema Nagarajan, "Study says Jarawas want to be left alone," *Hindustan Times,* Lucknow, 5th April, 2004, p. 11.
56. Sacred Space, *The Times of India,* New Delhi, 27th August, 2002.

REFERENCES

- Alvi, Seema, *The Hindu folio,* June 4, 2000.
- Auboyer, Jeannine, *Daily life in Ancient India,* (translated from French by Simon Watson Taylor), Asia Publishing House, London, 1965.
- *Bhagavad Gita,* Gita Press, Gorakhapur.
- Banerji, S. C., *A Companion to Ddharmasastra,* D. K. Printworld, New Delhi, 1998.
- Basham, A. L., *The Wonder That Was India,* Rupa & Co., Delhi, 1995.
- Chapman, J. L. & Reiss, M. J., *Ecology Principles and Application,* Cambridge Low Price Edition, 1995.
- *Cultural Heritage of India,* Vol. I, II, III, Ramakrishna Mission, Calcutta 1972.
- Das, Manoj, "Of Myths and Legends", *The Hindu folio,* 1 July, 2001.
- Dwivedi, O. P., *Environmental Crisis and Hindu Religion,* Gitanjali Publishing House, Delhi, 1987).
- Guha, Ramachandra & Martinez-Alier, J., *Varieties of Environmentalism: Essays North and South,* OUP, Delhi, 1998.
- Gupta, Dr. Narmada Prasad, "प्रदूषण की समस्या और लोकसंस्कृति", गगना चल, Oct-Dec 1999.
- Hoskote, Ranjit, "Looking for the River," *The Hindu folio,* July 1, 2001.
- *Hindustan Times,* New Delhi, March 23, 2001.
- "Is Our Forest Cover up ?", *The Times of India,* New Delhi, April 24, 2003.
- *India Today,* 22 April, 2002.
- Jahnavi, T., Sacred Groves of Manchar", *Geography and You,* Vol. 4, No. 1 & 2, Jan-Feb 2004, pp. 32-33.
- Jain, R. B., *Public Administration in India-21st Century Challenges for Good Governance,* New Delhi, 2002.
- Krishna, Nanditha, "Retreat of the Gods", *The Hindu folio,* June 4, 2000.
- Kumar, Sanjeev, *A Socio-Religious History of Daksina Kosala: A Study Based on Inscriptions,* M. Phil Dissertation (unpublished), JNU, Delhi, 1994.
- Kalki, Krishna, "पर्यावरण चेतना का प्रतीक ग्रामपर्व वट सावित्री", कुरुक्षेत्र, May 2004.
- Mitra Dr. Joytir, "Some Current Medicinal Plants of Vedic Antiquity", *Amrit,* Vol. I, No. 1, Jan-March 1988.

- Nair, S. M., "Rich Pockets of Green", *The Hindu folio,* June 4, 2000.
- Nagarajan, Rema, "Study Says Jarawas Want to be Left Alone," *Hindustan Times,* Lucknow, 5th April, 2004.
- Prime, Ranchor, *Hinduism and Ecology,* Motilal Banarsidass, New Delhi, 1996 (reprint).
- *Paryavaran Chetna,* September 2001 & Jan-Feb 2003.
- Pandey, V. B., "Evaluation of Medicinal Plants and Plant Base Medicaments," *Amrit,* Vol. I, No. 1, Jan-March 1988.
- Rastogi, Vandana, प्राचीन भारत में पर्यावरण-चिन्तन (विशेष सन्दर्भ : महाकवि कालिदास), Publication Scheme, Jaipur, 2000.
- Ravindra, Ravi, *Science and Spirit,* Paragon House, New York, 1991.
- Sacred Space, *The Times of India,* New Delhi.
- Sircar, D. C. *(ed.), Select Inscriptions,* Vol. II, Motilal Banarsidass, Delhi, 1983.
- Smith, Brian K., "Classifying Animals and Humans in Ancient India", *Man* (N. S.). Vol. 26, 1991, pp. 527-548.
- Srimad Bhagavatam.
- *Taittiriya Aranyaka of Yajurveda.*
- Thapar, Romila, *Asoka and the Decline of the Mauryas,* Oxford University Press, Delhi, 1973 (reprint).
- Varadpande, M. L., "Conserve water, says Mahabharata", *The Hindustan Times,* Delhi, 3 July, 2000.
- *Wake up India,* Jan-March 2002 & Jan-March 2003.

4

Environmental Ethics

Prof. A.N. Tripathi*

The dominant environmental issue of the day is pollution: pollution of air, water and soil. These three constitute the basic life support system for all living beings and vegetation on earth. A threat to them is a threat to our health, well-being and to even our continued existence as a species. This realisation has prompted several action programmes for fighting the menace of pollution. Scores of anti-pollution laws have been enacted by legislatures. Many regulatory bodies, like the Pollution Control Board, have been set up. There has been forceful judicial activism as well. The Supreme Court has come down heavily on polluters. It has also pulled up governmental agencies for their laxity. Another heartening development is the growing number of NGO's taking up environmental issues. A shining example is the forceful activism of late Sri Anil Agrawal. The Centre for Science and Environment founded by him, and the magazine *Down to Earth* brought out by the Centre, are doing a wonderful job fighting for environmental causes and spreading awareness about them amongst the general public.

Apart from pollution, environmental concerns have many other dimensions. The forest cover over earth's surface is diminishing rapidly. This causes floods, soil erosion, and environmental and ecological imbalances. The global temperature is rising due to the

* **Professor, Electrical Engineering, Institute of Technology, Banaras Hindu University, Varanasi.**

green house effect, caused by the increasing level of carbon dioxide in the atmosphere. The rise in temperature causes climatic imbalances. It also melts the polar ice cap raising sea level which threatens to inundate many low lying coastal areas around the globe. The protective ozone layer in the outer atmosphere is depleting because of release of Chloro-flouro-carbon gases used in refrigeration. This depletion means greater amounts of ultra violet rays in sunlight are reaching the surface of earth causing skin cancer and other diseases. Yet another threat is posed by disposal of toxic and nuclear wastes. They create long term hazards to human, animal and plant life.

All these problems have been created mainly by the modern materialistic and consumeristic attitude toward life. This lifestyle generates ever increasing demands for material goods and services. These demands have been met by rapid industrialisation using modern technology. The techno-economic forces aiding and promoting this quest have little regard for nature and environment. For them nature is merely a resource for industrial production. Its other utility is as a dumping ground for industrial and urban wastes. In this utilitarian view nature has only economic worth. It has no intrinsic worth, no rights of its own, and can make no moral demands on humans. It may have some aesthetic value. But that too only to the extent it serves the human need for aesthetic pleasures. This mercenary attitude toward nature was never seriously questioned in the heartland of modern industry and technology. This is because the dominant Western worldview considers man as the most privileged being in creation. He has dominion over the whole of nature; animals, plants, earth, seas, mountains, forests, rivers, lakes. They have all been created for the unfettered use of man, and are at his beck and call. In this homocentric view man has no moral responsibility towards nature. Fundamental ethical canons of moral behaviour, like 'do no-harm', and 'do good', apply only to our conduct towards fellow human beings, and not towards nature, plants and other non-human species. In fact nature was treated as an adversary, to be conquered, to be tamed, and to be exploited. 'Exploitation of natural resources' is a well respected phrase in the vocabulary of modern economists and technologists.

In sharp contract to this Western view, the Indian world view regards man merely as one of the many children of nature. He certainly has special qualities and attributes, but they do not give him power to despoil nature. At the spiritual level, the Indian tradition sees the same divinity in all living beings, and even in non-living things. They are all different manifestation of the same cosmic power Love and reverence for all life and nature are the cardinal tenets of this worldview. Living a life of self-restraint, frugality and simplicity, so as to make the least demand on nature and cause it the least injury, are the central values of its admired way of life. But the fundamental questions of environmental ethics, namely what is our moral responsibility towards nature ?, and what ethical principles ought to govern our interaction with nature?, have not been raised forcefully enough in our country as well, and certainly not after Mahatma Gandhi. To him reckless exploitation of nature was against Ahimsa, the highest ethical principle regulating our conduct towards all, including nature. He proposed a model of development based on traditional technologies, cottage industries, and self sufficient villages as the basic unit of economy activity. In this model production was distributed and not centralised; it relied more on human energies and human capacities; and the production was for consumption not for profit. It was much better suited to Indian social conditions, and much more eco-friendly than the modern energy and capital intensive centralised methods of production.

Yet, the Gandhian model for socio-economic development was given a go by after Independence. The ruling elite of the time chose the Western model of large scale industrialisation for the reconstruction of national economy. To promised much faster growth to meet the pressing need for the basic necessities of life; food, clothing, shelter, health care, education, transport, communication, etc., for the poor and the downtrodden masses. It also had the glamour of being modern. Once unrolled the juggernut of modern development steadily gained momentum, riding roughshod over all ideological and environmental objections. It has accelerated still more in the recent times, goaded by the battle cries of globalisation, privatisation, competitiveness etc. It has enthusiastic support of the elite and upper middle classes, as it satisfies their consumersitc cravings for obtaining 'international' standards of conveniences and luxuries. And in this pursuit they are as unmindful of the social and environmental costs as their counterparts in the West. The respectful and reverential sayings of the Indian worldview towards nature, animals, plants, rivers, mountains etc., are used only for platitudinous references, and not for guiding individual conduct or public policy.

The current global concern for nature and environment does not arise from any ethical commitment to honour the rights of nature, and to protect its sanctity and integrity. Rather, it stems from the utilitarian motive of protecting the interests of humans. It is only when our exploitation of nature has rebounded to pose a threat to our basic life support systems that we have become environmentally active. In the ethical sense it is a kind of selfishness, not so much of the individuals but of the human species as a whole. Discussions on the ethical dimensions of this environmental activism are mostly confined within the paradigm of human-centered, utilitarian ethics. But in spite of its theoretical limitations this paradigm is not very restrictive in practice. It has generated a new awareness about our relationship with nature and thereby contributed to positive measures for protecting nature and environment. This has been called environmentalism or eco-wisdom.

One contribution of this new awareness is the concept of our moral responsibility towards future generations. If we continue to exhaust non-renewable resources and continue to degrade the quality of environment, the future generations will have little left to meet their requirements for sustaining a good quality of life. Is it morally right to deprive the coming generations of their right to fulfil their needs of survival, safety and enjoyment of life ? The ethics of responsibility demands that we feel concern for the well-being of the whole of humanity; not only for those of our fellow beings here and now, but also for those who are not yet born. This ethics has prompted the new concept of 'sustainable development', which has been defined as "...the development that meets the needs of the present without compromising the ability of the future generations to meet their own needs" (Report of the World Commission on Environment and Development). This ethical perception must inform all our individual choices, as well as all our environmental and developmental policies. The burden of acting according to this enlarge concept of ethics of responsibility falls much more on those who exercise modern forms of powers; political, economic, administrative or professional. Their actions are in the public domain, in the form of policies and there implementation. They should ever be conscious to carefully examine the impact of their actions and decisions on nature and environment. In their efforts to seek short term gains for the constituencies they serve, they should not jeopardise long term interests of the humanity itself. This ethics of responsibility needs to be further developed into more specific areas. It should cover not only actions of individuals but also those of governments, corporations, and all other social institutions and organisations.

The other significant contribution is the evolving concepts of deep ecological awareness. It transcends the narrow confines of human centered utilitarian ethics, and comes very much closer to the spirituality of traditional Indian worldview. Capra calls it. "the intuitive awareness of all life, the interdependence of its multiple manifestations and its cycles of change and transformation" (*Uncommon wisdom,* p. 245). Some other modern philosophers enlarge this intuitive awareness of oneness to include not only all life forms but even insentient things like mountains, rivers, landscapes etc. Some of the principles of this new ecological ethic, as noted by Peter Singer (*Practical Ethics,* p. 281) are :

1. The well-being and flourishing of human and non human life on earth have value in themselves (synonyms: intrinsic value, inherent value). These values are independent of the usefulness of the non-human world for human purpose;
2. Richness and diversity of life forms contribute to the realisation of these values and are also values in themselves;
3. Human have no right to reduce this richness and diversity except to satisfy *vital* needs.

The worldview reflected in this deep ecological awareness is a radical departure from the traditional western worldview. And although it appears very much congruent with the Indian view, there are many basic metaphysical differences between the two.

However, the important thing is not to probe the philosophical niceties, put to examine the practical implications of this new ecological ethics. But, before doing that let us take note of another ethical principle related to exploitation of nature, and pollution of environment. This is the principle of equity, justice and fairness. It is being grossly violated, both at the international and at the national levels.

At the international level it is the economically and politically more powerful nations of Europe and North America who have been the main beneficiaries of modern techno-economic development. They are the biggest exploitators of nature as well. The United State o America, which has only six per cent of world's population, uses nearly fifty percent of the earth's resources. This is grossly unjust. The developed Western countries are also the greatest polluters of environment. But the harmful consequences of their greediness and environmental barbarity have to borne by all the nation and communities of the world. The poorer nations of the world are ill equipped to bear this brunt as they lack financial and technological resources to protect themselves against the environmental onslaught. And the West is in no hurry to share its know-how with them. Western nations have cut off most of their forests, but they are forcing poorer nations to protect theirs, so as to cleanse the air polluted by them. They are dumping their toxic chemical wastes and far more hazardous nuclear wastes in third world countries or in international seas. A glaring example of callousness and ethical irresponsibility of this kind is the USA's recent refusal to ratify an international treaty, the Kyoto pact, which will require it to reduce its greenhouse gas emission.

A similar pattern of inequitous distribution of the benefit and burdens of techno-economic development is clearly noticable at the national level also. While the benefits go to the well-off urban classes, the burdens are passed on the poor and disempowered rural communities and adivasis. It is the livelihood and the living space of the latter which are submerged by a high rise dam, while the electricity generated by it goes to run the air conditioners of elite city dwellers. The Pollutants produced by the industrial-urban sectors poison the rivers, lakes, soil which are the lifeline for agricultural communities. While the rich can safeguard their health by drinking bottled mineral water, the villagers and their cattle die of drinking polluted water. Thus, while the benefits of techno-economic development go the rich and the elite, its environment costs are passed on to the poor and the disempowered.

All the above noted issues concerning nature and environment need to be examined in depth as important problems of the day. Such a theoretical examinations will bring out the fuzzy and hidden dimensions of the problems and thereby help clarify our ideas for advancing the concepts of environmental ethics. However, at every step of this advancement we should also pause to examine the operational implications of the ethical ideas and principles and the constraints on putting them into practice. The theoretical and the practical considerations should reinforce each other. And it is this interdependent advancement of theory and practice which constitutes the real advancement of civilisations.

The first and the foremost implication of the philosophy of deep ecology and other components of environmental ethics would be to put a question mark on the modern materialistic, consumeristic and individualist philosophy of life. This ruling ideology and the life styles based on it, have been challenged in the past by the traditional idealistic, spiritual and religious views of life. However, these challenges have proved largely ineffective. This is because they have been based on abstract, metaphysical, and often less than rational ideas and belief systems. They could not contain the natural human urge to improve the material quality of life to make it more congenial, comfortable and enjoyable. The main tool in this enthusiastic quest was the new found power of modern sciences and technology. In due course this urge took the shape of unfettered hedonism, selfish and exploitative individualism, and an amoral if not immoral view of life. The earth and its resource were considered inexhaustible to meet the endless demands of human needs, greeds an lusts. And also that the atmosphere and the biosphere had infinite capacity to absorb and to purify the industrial - urban wastes, effluents and other pollutants. The industrial, technological, economic and political systems thrived on this myopic view.

The falsity of this view started becoming apparent after the middle of the last country. The acute problems caused by pollute atmosphere, poisoned rivers, denuded hills, plundered nature and exhausted resources were felt by every one, particularly by the highly industrialised West. Publication of expert studies, like the club of Rome report, *Limits to Growth,* 1972 and the report of the World Commission on Environment and Development, *Our Common Future,*1987 left no one in doubt about the seriousness of the threat to human well-being caused by environmental degradation and depletion of resources. The challenges to hedonistic philosophy of life posed by this threat was very real, concrete and visible to all.

The response to this challenge was at two levels; one practical and the other conceptual or philosophical. Several positive actions were initiated at the practical levels; stricter laws to check pollution, environmental preaudit of all development projects, conservations of non-renewable resources, stress on sustainable development, use of eco-friendly techonologies, recycling of waste, international agreements to control pollution and promote conservation, etc.

At the conceptual level there is greater awareness of the mutuality and interdependence of the nature-human relationship. The need to protect, preserve and nature plant and animal life is becoming more and more internalised in the common psyche. There is greater respect for the wisdom of traditional communitarian life styles which have a much greater degree of harmony and balance with nature. The philosophy of deep ecological thinking is slowly but surely gaining greater acceptance in the intellectual temper of the modern times. The concepts of environmental ethics are being discussed and published. It is well on its way to becoming an important sub-discipline of applied ethics. This development needs to be pursued much more vigorously in our country, taking into account the Indian socio-political realities, and the Indian world view. The principles of environmental ethics should be based on, and should emerge out of this world view. It will then find greater resonance in the minds and hearts of Indians. Furthermore, they should not be left at the level of abstract generalisations but should be developed into sufficient details so as provide guidance for human conduct, both at the personal level and at the level of public policy. Then only this new ethic will acquire sufficient strength and acceptability to positively impact the course of socio-economic development in a way that satisfies both the moral sensibilities and the human aspirations for a better quality of life.

5

Environment in Contemporary Hindi Poetry

Dr. Suman Jain*
Dr. Niharika Lal**

What is life? It is an amalgamation of earth, water, fire, air, vegetation and the senses. As has been said by Tulsidas -"*Chiti jala, pāvata, gagana, samīrā, pañca tattva se racita śarīra.*" That which is at the core, is perceptible in the entire universe. This can be seen in the childhood pranks of lord Krishna. When Krishna says that he has not eaten clay, his mother, Yashoda, asks him to open his mouth, and perceives he entire universe inside it. When one becomes aware of the life-force pervading the entire universe, then only can one have this experience. The life- cycle of the world proceeds on the basis of plants, bacteria and other animate and inanimate objects. It is through a biological process that we get the edibles, coal, oil and minerals. The maxim "parasparopagraho *gīvānām*" indicates that all forms of life are interdependent. While scientists hold that all creation is inanimate, theologians believe that it is animate. Like human beings, the earth, water, fire, air and vegetation, too, are animate. This animate consciousness give birth to various substance. Indian thought gives so much importance to consciousness in each object of nature that even the mountains Vindhya and Himalayas, the rivers Ganga and Yamuna as well as

* Senior Lecturer, Department of Hindi, Vasant Kanya Mahavidyalaya, Kamachha, Varanasi.

** Senior Lecturer, Department of English, Vasant Kanya Mahavidyalaya, Kamachha, Varanasi.

the waves of the oceans are referred to in our national anthem. The tiger, peacock and lotus are the national animal, bird and flower respectively. Horses, bulls and lion find a place alongwith the inscription "Satyameva Jayeta" on the national emblem. The entire populace, from the Himalayas to the Indian ocean, consider the plants, animals, birds, fish etc. to be a part of their family worshipping the plants and trees of Tulsi, Pipal, giving food to the birds, taking out a morsel of food for the cow before partaking it, looking upon the cow, the river Ganga, the country and earth as venerable as one's own mother is the sacred duty of each citizen. It is considered a sacrilege to cut trees, pollute water, torture birds and animals and to misuse or waste anything. Appropriate use of foodgrains, clothes and time leads to the enhancement of community life and an environmental balance is maintained. One who has accomplished this sees his own reflection not only in each of the 84 lakhs of forms of life, but in each object of the world and its creator. The various incarnations - Matstyāvatara, Kacchapāvatara, Shūkarāvatara, Nrisimhāvatāra, Vāmanāvatāra—are symbolical of the expansion of our family. Undeniably, the Indian belief is that:

> "Vyāpaka visvarīpa bhagwāhā, tehi dhari deha carita krita nānā."

After studying the intensive relationship between man and the rest of the animate as well as inanimate world, let us now turn our attention to the fact that there is such a comprehensive interaction between nature and society that it influences the entire mankind and gives rise to what is known as the environmental question. Continuing scientific and technological progress has provided man with immense powers over nature. Today, man is capable of removing mountains and changing the course of rivers. We can bring about original changes in nature. Our productions as well as other economic, scientific and technological activities are spread from the earth to space and everywhere man is exploiting nature. According to the life Surrey of India, in the year 2003, due to the climatic and environmental changes, only 89, 451 varieties of creatures and 47,000 species of trees and plants are to be found in the country. Blind and thoughtless destruction of jungles has led to the extinction of nearly 1336 species. Due to the extensive exploitation of land, pollution, use of pesticides, floods, famines, storms, landslides and other natural calamities, there is a possibility that 39 species of mammals, 72 of birds, 77 of reptiles, 3 of amphibians, 2 of fish and thousands of butterflies and insects will go extinct because the productivity of the earth is decreasing day by day.

When we study environment in the light of literary works, we find that our ancient texts like the *Vedas, Upanishads, Puranas,* etc. contain fundamental esoteric formulae and objectives related to environmental education and studies. During the *Vedic* age, the sages propounded great philosophical thought, residing in the natural surroundings of the forests. They expatiated on the fundamental questions related to creation, life and the world. The *Vedic* culture was the culture of forests. In that age, the hermitages were the centres of learning and education. It was in the lap of nature that the excellent *Vedic* philosophy, culture and science flourished. The Aryans propagated the worship of all components of nature viz. the Sun, water, air, earth animals, birds, and vegetation so that they would be compatible and calm towards mankind i.e. they would be available in a balanced way. The Aryan sages allotted a deity to each element of nature—Agni to light and warmth Sun to energy, Varuna to water, Dharti to earth or clay, Maruta to air, Ashwini Kumars to herbs, etc. The *vedas* contain progress in order to make the above-mentioned elements favourable and useful to mankind. There was a custom amongst Aryans to pray for peace at the beginning of each function, a custom that is prevalent even today. The *Vedic* sages prayed:

> "aum dyanḥ śantirantariksa - aśāntiḥ pṛthivi śanti rāpaḥ
>
> śānti roṣadhayaḥ śānti: vanaspatayaḥ śāntirviśvedevāḥ
>
> śāntir brahmā śāntih sarva śāntiḥ śānti deva śānti:
>
> sā mā śānti redhi

Hence, it is clear that the *Vedic* seer endeavoured to protect and maintain a balance amongst the fundamental elements of nature. The seers knew that the Sun was the source of energy. The light and warmth of the sun infuses life in the creatures and vegetation of the earth:

"Divyam suvarnaṁ Vāyasam brihantamayām garbha darśanamoṣadhīnām.

Abhīpato vrrsṭibhastapayantam sarasvantamavase johavimi"[1].

i.e. we invoke the God Sūrya one who is force behind the fast moving, glorious waters to be found on earth, who nourishes and strengthens the medicinal herbs and who satisfies the earth through rains, to provide protection to us.

The earth is the progenitor of food, herbs and vegetation as well as precious metals and jewels. Man gets nourishment through grains and vegetation. Hence, the earth is conceived as a mother and enlogised as such in the Atharvaveda:

"mātā bhūmiḥ putro aham pṛthivyāh"[2].

The sages hold that one should not destroy either trees or other forms of vegetation. "Svadhitai me na hinsi"[3] i.e. don't cut the trees with an axe. The purāṇas also state that trees should not be cut:

"Aśvattho vatvrkṣaśca na chennagyah kadācana"

Great importance is given to the basil plant (tulsi) in the purāṇas. Regularly consumption of basil leaves leads to good health. There is a discussion on the importance of the basil plant in the padmapurāṇa. Lord Shiva says that among all the leaves and flowers, the basic leaf is the best. Its use leads to a rise in fortune, fame, glory and knowledge and a man becomes healthy and free from disease:

"sarvebhyaḥ patrapuṣpebhyaḥ sattama tulasīsivā

sarvakāya pradā śuddha vaiṣṇavī viṣṇu supriyā."

It is believed that modern environmental thought came into existence with the naturalistic philosophy of Jean Jacques Roussean. According to Rousseau, nature in its original form is pure and beneficial. It is becoming polluted because man is disturbing it. He gave the slogan - "Return Back to Nature."

According to Acharya Ramchandra Shukla, a renowned critic and historian of modern Hindi literature - "Man destroys his own happiness by severing his ties with the rest of nature. Just as man finds immense space and scope for the expansion of his intellect. Similarly he finds it for the pervasiveness of emotions or feelings also. But if due to laziness or negligence. He limits the latter, then in no way can his happiness be said to be superior to that of animals. Thus it is proved that our love for forests, mountains, rivers. waterfalls, animals, birds, fields, groves, etc. is natural or at least is inherent in our conscience in the form of desire or longing."

The prime objective of poetry is to assimilate both the animate and inanimate and make the readers experience them. In order to achieve this objective, it is essential that one should sacrifice one's pride without doing so, one cannot experience all the forms of nature. This completeness of nature is discernible in the writings of Sanskrit poets. In Sanskrit literature, nature is portrayed as the heroine. If we remove nature from "*Abhijñāna-Śakuntalam*" and "Meghadūta", then these will cease to be literary works. Kalidasa's description of nature is natural. The poet involuntarily describes the natural beauty in minute detail:

"Sa palvalottīrṇavārāhayūthā yāvāsavṛkṣonmukha barhiṇṁ

Yayaū mrgādhyasitaśadvadatāni śyāmayamānāni klāni paśyan"[5].

King Dilip surreys those green forests were wild boars were coming out of small pools, peacocks were returning to their abode in the trees, and deer were sitting on the grass. Similarly, the first poet, Vālmīki, has not only graphically described the mangoes surrounded by sprouts, creeps laden with sweet scented flowers and lotuses laden with pollen and fragrance but has also minutely portrayed wild trees, trees and plants like acacia, ingudi, anikota, tendu, baheda etc.

"Mountains, forests, rivers, waterfalls and other natural scenes are generating sources of emotions independent of feelings of passion or conjugal love. They naturally attract warm-hearted people as well as literatures. Our disposition becomes engrossed on getting and inkling of the basic forms and circumstances of life included in the objects and activities of these scenes."[6]

The compositions of Bharatendu Harishchandra, the originater of modern Hindi, are relevant even today:

"nava ujjvala jaladhara hāra hiraka si sohati

bica bica chaharati boond madhya muktā mani pohati."

Today, the condition of the river Ganga is extremely pathetic. In the context of Ganga, the statement of Sunderlal Bahuguna, is extremely relevant — "The greatest danger to the existence of Ganga is the Tehri dam. After the construction of the dam, the soil flowing in the Ganga during the rainy season, will not be able to make the plains of Uttar Pradesh fertile, as it will be deposited in small pools. During the rainy season, the floods used to carry the soil to the fields via the canals and this was a boon to the poor people residing on the banks of the river as they could grow vegetables. The level of underground water was also maintained. Even the imagination of that great disaster, which has been predicted by scientists due to the construction of the dam on the land prove to earthquakes, is extremely for frightening. If the dam breaks, then Rishikesh will be under 260 metres of water in 63 minutes and in the next 20 minutes Haridwar will be under 232 metres of water. After 12 hours the water will reach Bulandshahar destroying Bijnor, Meetut, Hapur, etc. on the way."[7]

Similarly the river, Yamunā, is described in "Chandrawali"

"surabhi tanuja taṭā tamāla taruvara bahu chāye

jhuke kūla son jalaparasana hita manhun suhāye

kahim tīta par amala kamala sobhita bahu bhāntin

kahim saivālana madhya kumudinī lagi rahi pāntin"

Today the rush for materialistic gain has led to the destruction of the natural beauty of both Ganga and Yamuna. The construction of power-plants and densely populated cities on the banks of these rivers are destroying them. How nature revolves against man's exploitation is described in "Kamayani" of the great poet, Jaishankar Prasad. The first cants of Kāmāyani is the "Cinta Sarga" and in it nature takes on a terrifying form due to the unrestrained behaviour and unrestricted and self appeasement of the gods. As a result a deluge takes place:

"saurabh se digant pūrita tha, antariksha āloka-adhira,

sabmein ek achetana gati thi, jis se pichara raha samira

sura surabhimaya badan aruna the nayan bhare ālas anurag,

kal kapol tha jahan bichalta kalpavriksha ka pita parāg

vikal vasana ke pratiñidhi ve sab murjhāye chale gaye

āh! jale apni jwala se phir ve jal meinjale, gaye.

By comparing the epic, Kāmāyani, Jaishankar Prasad has proved that unless there is a mixture of balance, desire, action and knowledge in order to preserve the beauty and expansion of nature, the development of mankind is impossible.

Artists and the human mind have understood both the frightening and beneficial forms of nature. Romantic as well as contemporary poets have portrayed nature in an innate manner and with an environmental viewpoint.

Nature is always present in the love-lyrics of Agyeya. It is not in the form of an excuse or screen but intensifies the association of love even further:

"bhale ki ani-si bani bagulon ki dār

phulakiyān chitput gol bandh dolatin,

sihran uthati hai ek deha mein

koi ko padhārā nahin, mere sune geha' mein tum phir aagayāe

Kwāra"

Nature, in Agyeya's poetry, is easily assimilated. The earth harmony of words can be seen in "tum phir, aa gaye kwara".

Trilochan Shastri is a poet deeply rooted to the early. In his poems, the frolics and beauty of nature are depicted. His views about nature is very much similar to that of an Indian farmer who gets fascinated on perceiving small plants growing amidst the greenery of nature. Like Nirālā, Trilochan has also depicted many scenes of monsoon. While doing so, it appears that he is not trying to create remarkable and extraordinary scenes of beauty but is trying to invoke the spirit of the farmer lying somewhere in his consciousness. He express his innate love for the land, crops and soil:

"ulha kisan o, ulha kisan o, badal ghir aaye"

Nature, life and farmers not only give him a separate identity but also provide a clue to understand his world-view

"us janpada ka kavi hun jo bhūkhā dukha hai

nanga hai anjaan hai, kala — nahin janta

***

jhānya jhānya karti dupahariya nacha rahi thi,

jalti hui. Bhaur si garmi ki pagdandi

mujhe le gayi amon ki bārī mein ki thi

nahin adhik ki asha. Pā kar chāyā thandi

ankha mund kar socha man mein, swarga yahi hai.[8]

The poetic sensibility of Nagarjuna belongs to folk life. This is clear from Reway, he describes the dew laden milky sprouts of paddy in Agahana (the ninth month of the Hindu calendar). He exuberatedy states that his heat lights up when two roasted ears of corn are placed before him. He expresses his feelings in the following words — "subah subah / gavai alāva ke nikat / ghere me baiture— batiyāne ka sukh lootā / subah subah / āñcalika beliyon ka murture / kānon ki in kaṭoriyon mein bhar kar lauṭa / subah subah / sahar aur kolahal bhare jeevan se ūbakar kavi apne goan ki or lauṭta hai to abhiryakti apni sundartā ke saath vyatla hoti hai—bahut dinon ke baad / ab mein ji bhar sun pāyā / dhām kūtati kishoriyon ki kokila kanṭhi tana / x x x x / ab ki ji bhar choo pāyā / apni ganvayi pagdaṇḍi ki chandanvaṇī dhul bahut dinon ke baad." Only the branches of the neem tree provide him with solace in the fast life of the city, rest with the noises of machines:

"bethabar soi hui hai chāpne kī yah virāt machine

udhar munh baye pade hain taipon ke malin dhisar case

par, idhar to jhanktī hai do sālonī ṭahaniyān sīkhchon ke paar."

The poetry of Sarvashwar, a poet of the masses, breaks away from the myth of the classics and adopts the style of the common people. In the poem, "*Rishta*", the poet analyses the relationship between the tree and the bird and states that this relationship is eternal:

"kitnā nirāla rishtā hai/peḍa ka chiriya se/jise wah apna mānta hai/aur jiske sahare/har baar apne ko/naye sire se pehchantā hai."

In the contemporary poetic scene, Kedarnath Singh is a poet who belongs equally to both the village and the city. The realities of Indian culture and creative facilities are unified in his poems. In order to understand his poems, it is not necessary to refer to the classics, rather one should go directly to the villages, markets, fields, meadows etc. where his poems are born, nurtured and decorated. "Kedar's poems are not a reflection of reality; they always point to the turmoil lying at the base of reality. In his poems, one hears the voices of crops refusing to go to the storehouses or markets and the tearful eyes of the fish are depicted, expressing a desire to live." The poet so points to the silence speading over the city. There is no place for scientists and the co-existence of men is coming to an end:

"is itne baḍe shahar mein

kahīn rahtā hai ek kavi

wóh rahtā hai jaise kunye mein rehti hai chuppī

jaise chupp main rehte hain shabda

***

woh rehta hai is itne bade shahar mein aur kabhi kuch nahin kehtā

aise parivesha mein woh apnā rāstā bhī ḍhunḍhā liyā hai

ab drishya bilkul sāf thā

as hamare samne

gāya thī

kisan thā

rasta tha

sirf hamīn bhool gaye the jana kidhar hai."

The solutions to current environmental problems can be clearly found in his poems. Multinational companies are taking over the villages and markets of India, but the poet preserves them in his writings. Even though he wants to, he cannot break a branch of the neem tree:

aur chhoḍ dī neem ki kahanī/phir laut āyā/rāste bhar khud se ladtā aur khud ko samjhota huā/ki yadi kahin toda hi letā/kuch hari kacchi ubharti hui shākhen/to uska kyā hotā/jo unhin ke niche baiṭḥā thā/unhīn ke jādū mein/banda kiya ankhen."

Similarly, there is another poem of his entitled "*Daana,*" which provides a solution to the problem:

"nahīn, hum mandī nahin jayenge/khalihan se ulhate hue/ kehte hain dāne."

The poem, "*Pakshi ki vāpasi*" written in 1983 reflects a picture of today's environment:

"kyā nāme hai uska/khanjana/ṭiṭihari/nīlakantha/mujhe kuch bhi yaad nahin. Mein kitni āsānī se bhooltā jā rahā hoon/ pakshhiyon ke naam/mujhe soch kar dar lagā. Birds and animals, insects etc. are very essential to us and are a part of our lives. Today, the situation is terrifying. For the past few years, there is talk of eagles and vultures going extinct who earlier were to be seen in huge numbers over the vast area of North India. We used to see the eagles and vultures scaling the sky or feeding on the corpes of animals in jungles and fields in huge numbers but has their number is diminishing. The reason for this was not yet ascertained when suddenly the crows—whom we were habitual of seeing on our windows morning, afternoon and evening started disappearing. Now, no more do we hear the cawing of of crows at dawn. We used to see crows fighting each other over a piece of bread or any other morsel of food, but now this scene is becoming rare. The crows are leaving us. They were a part of Indian culture and literature. The Maithile poet, Vidyapati, has described the crow in the following words:

"more re anganwān chanan kerī gachiyā

tāhi ūpar kurai rī kāga

sone choncha maḍha deba tohe hum kāgā

Jehi ahirya piyawā hamāra

The heroin is waiting for her lover who has gone to foreign lands. She is agitated and frustrated. At this moment she perceives a hope, belief and good omen in a crow. There is a sandal tree in the heroine's courtyard and a crow comes and sits on it. The heroine says that O crow! If my lover returns from foreign lands, then I will mount your beak with gold.

Similarly, Jāyasī's heroine Nāgamati says— "Kāgā sab tum khaiye—chun chun khaiyo masa/do naina mat khaiyo tha pura piya milan ki aas."

According to scientists, the reason who the number of birds is diminishing is that man is blindly using pesticides in order to enhance the yield or preserve it and this is proving to be a danger to the birds. Birds and animals instinctively realise the dangers of pesticides and flee from the place. But now they are finding similar pollution at other places and hence, their species is slowly going extinct. In the view of biologists, birds end animals exist on the vegetation and flesh of other birds and animals and these are slowly becoming poisonous. We feed our pets at homes and the chemical fertilizers that we put in our fields is poisonous beyond our imagination. As a result, not only birds and animals but also human beings consume this slow poison. This is the reason why many pesticides and some chemical fertilizers are banned in developed nations. The matter does not end here. The traders of those nations are exporting those pesticides and fertilizers to developing countries thereby earning a huge profit.

It is a fact that the environment today has become poisonous. The way poisonous substances are being consumed and mixed at the villages, towns, cities, cosmopolitan cities, jungles, mountains, waterfalls, rivers and oceans has become a source of worry for the very existence of life. And the responsibility lies in human life. In the blind race for development, man is forgetting that he is cutting the very branch on which he is sitting.

On going through the literature of our country, one fact becomes very clear to us. Nature or environment, which figured as the heroine in earlier literary works is becoming a source of worry in contemporary literature. The poet and artists is agonised over this terrifying fact. Again and again, they are reflecting on the negative and positive aspects of this problem. Today, wastage of water, cutting down of forests, killing of birds and animals, destruction of vegetation, digging and exploitation of land is quickly polluting the environment. Hence, it is essential to impart education dealing with the necessity of maintaining a balance in life along with that which is imparted for earning a livelihood. Imbalance leads to problems not only in family life but also gives rise to crime. We are individuals but at the same time we are members of a society. No one can be completely devoid of selfishness, but it should be ascertained that one's desires do not conflict with that of others. In a society or community a restrain must be placed on desires or self-interest. If such a restrain is kept then the problem of pollution will never arise. Nature reveals her magnificence to the energetic Lord Shri Krishna hints towards this when he says:

"paśya mein pārtha rūpāṇi śatāśo atha sahastrah"

i.e. perceive in the flowers, fruits, leaves, buds, plants and also in birds, animals, inanimate, creatures, world. For those who are absorbed or engrossed in the Almighty, this is spirituality or the vision of the Almighty, self-revelation for those immensed in themselves and environmental thought or philosophy for students.

REFERENCES

1. *Ṛgveda,* I, 164, 52.
2. *Atharvaveda,* 12.1.12.
3. *Yajurveda,* 4.1.
4. Acharya Ramchandra Shukla, *Chintāmaṇi,* Part 2, p. 4.
5. *Raghuvamśa,* 2.17.
6. Acharya Ramchandra Shukla, *Chintāmani,* Part 2, p. 29.
7. Jane Jana ki gangā maiyā. p. 3.
8. Trilochan Shastri, "Vs. janpada kā kavi hun."
9. Kedarnath Singh, *Pratinidhi kavitāyen,* p. 7.

6

Environmental Education and Its Concern in Educational Policies in Independent India

Dr. Sunil Kumar Singh*

According to Earth Matters (6.3.2001) level of Green House Gases (Carbon-di-Oxide, Methane, Ozone etc.) especially Carbon-di-Oxide has increased tremendously in the last 140 years only, and by the end of the century it is expected to double. In the last century 2^0C increase in temperature has already been noted. Besides this, 2.5^0C increase in temperature in last 50 years in Antarctica has also been noted, which has resulted into unusual melting of a big block of ice in the continent" (National News, 8.30 p.m. 20.3.2002). The increasing temperature of Earth has also resulted into increase in heat waves, drying of water resources and so on. If warming of the Earth continues in the same manner, it will lead to series of chain reactions like fast melting of snow and glaciers, rising of sea level, submergence of low lands, unusual flooding of rivers, change in rainfall patterns, emergence of violent thunders/storm and change in climate leading to change in natural habitates. Change in habitates and life supporting temperature may lead to emergence of several diseases in epidemic form and ultimate loss of several species of organisms. For example, according to Earth Matters (6.03.2001), 70 per cent coral reef in India ocean has already died due to ocean warming, Penguin Population in Antarctica is decreasing due to melting of ice and survival of gentle giants of the ocean (Whale)

* Reader, Faculty of Education, Banaras Hindu University, Varanasi.

may be threatened. May be this is also one of the reasons that the well known natural scavangeres i.e. "Vultures" are out of sight today.

Besides the loss of species of organisms, the increase in complexity of organisms is another problem related to environmental changes. Revival of 'Plague' is most recent example related to this. The microbes, pests and insects are showing resistance to severval generally used pesticides and insecticides. All this may lead to revival of totally eradicated diseases in near future.

Not only the above but with the increasing temperature of environment due to unplanned, lustful and uncontrolled human activities in the name of human development, the water availability, cropland areas, forest areas are decreasing at a very fast rate. On the whole the attitude of human indifference towards the environment has resulted into pollution of fresh air, water and land etc.

A review and thinking over the above issues clearly reflect that all of them are indicaters of 'environmental crisis', the crisis related to survival of living organisms on the planet Earth. The challenge of environmental crisis can be met only by, (i) spreading environmental awareness; and (ii) initiating environment friendly activities, Both of these can be achieved through Environmental Education (EE).

Environmental Education

Environmental Education is an educational intervention to meet the challenge of environmental crisis. Below given chronology of environment related events reflect the historical emergence of Environmental Education at the international level.

The table *(See next page)* clearly shows the development of Environmental Education. According to Belgrade Conference (1975)the goals of EE are:

> "To develop a world population that is aware of, and concerned about the environment and its associated problems, and which has the knowledge, skills, attitudes, motivations and commitment to work individually and collectively towards solutions of current problems and the prevention of new ones".

Sl. No.	*Year*	*Place*	*Event*	*Result*
1.	1972	Stockholm (Sweeden)	United Nations Conference on the Human Environment	Establishment of United Nations Environmental Programmes (UNEF)*
2.	1975 (October)	Belgrade (Yugoslavia)	International Workshop on Environmental Education (Organised by I.EEP)	Belgrade Charter (contained framework, objectives and guiding principles of EE)
3.	1976 (November)	Bangkok (Thailand)	Asian Regional Meeting**	Recommendations for programmes for EE, personnel training; Non-formal E.E., and materials for EE
4.	1977 (October)	Tbilisi (Georgia)	Inter government Conference on EE	First phase of EE programme culminated

* UNEP alongwith UNESCO lanuched International Environmental Education Programme (IEEP).

** Besides this meeting, a series of meetings were organised in different continents Africa (Brazzavills), Arab States (Kuwait), Latin America & Carrabean (Bogata), and Europe (Helsinki).

Source: Source Book on Environmental Education for Elementary Teacher Educators (1994) — Table has been developed by author based on information obtained from the book. The goals and objectives of EE on page 4 & 5 are also based on it.

Further, the Tbilisi Conference (1977) extended it and developed the objective of EE. They are as follows:

- *Awareness:* to help individuals and social groups gain a variety of experience with the total environment and to acquire a basic understanding of the total environment, its associated problems and humanity's critical responsible presence and role in it;
- *Attitude*: to help individuals and social groups acquires social values for working towards the solution of environment problems and to faster a dialogue between these groups;
- *Skills:* to help individuals and social groups are the skill for working toward the solution of environment problems and to faster a dialogue between these groups; and
- *Participation:* to help individuals and social groups to develop a sense of responsibility and urgency regarding environmental problems to ensure appropriate action to help solve these problems.

According to Sharma (1994), after the culmination of first phase of development of EE a regional workshop of participants from Member states was held in Bangkok in Sept. 1980. Following recommendations were made during this workshop:

1. To stimulate and facilitate the exchange of information; to continue to enhance an inventory of national institutions and programmes in environmental education;
2. To continue to collect and disseminate information and documentation. The issue of news letters was considered as one of the means for this purpose;
3. To prepare and publish various source materials on environmental education for useing in curriculum development and teacher training at the national level, including monographs on environmentally important eco-systems such as paddy field, water reservoirs, coastal regions and river basins;

4. To organize study visits and attachments, including these specifically for exchanging experiences among youth leaders;
5. To provide support to studies, research and experiments at national level;
6. To provide advisory services to, and paritcipate in activities at national level;
7. To develop a pool of experienced resource persons in the region for their possible consultancy services to the Member states.

Later on in 1989 a Training Workshop in EE (1989) was organized for Association of South East Asian Nations. According to Sharma (1994) the objectives of EE identified at this workshop are to enable students to:

1. Acquires basic understanding of the interrelationship of components and factors of the environment and the processes that occur in it;
2. Develop desirable values and attitudes, especially concern and responsibility towards the conservation and enhancement of the environment;
3. Acquire refine skills in identifying, assessing, and solving environmental problems;
4. Develop a sense of urgency in responding to environmental issues and problems and taking action toward their solution.

The above sections clearly reveal the emergence, development, goals and objectives of EE at the international level in general and Asian region in particular. All the countries of the world have adopted EE either in the integrated form with other subjects or as a separate subject. Let us review the concern for EE in Education in independent India.

Concern for EE in Educational Policies in Independent India

Concern for environment is seen in reports of commissions, National policies and the curriculum frameworks published from time to time. Report of the Kothari Commission (1970) suggested that, "Science education at all levels should be strongly reinforced through study of applications to local environment and industry;" (p.720). The commission also suggested the following:

> "... subjects like geochemistry, geophysics and geomagnetism and economic geology are of great importance in the exploitation of the natural resources and should be emphasized in the training of geologists. In the field of biology, the study of micro-organisms and their role in medicine and agriculture deserves much more attention than is generally the case. Universities and colleges should make full educational use of museums, public botanical gardens and zoos and scientific and industrial institutions." (p. 735).

A special course on "Environmental Sanitation" (p. 740) was also suggested by the commission. In this way, the suggestions given by commission were not directly related to EE, but study of environment and general sanitation was certainly suggested. It reveals the concern of commission for general awareness about environment.

Based on the recommendations of Kothari Commisiion, National Policy on Education (NPE) 1968, was formulated. According to Aggrawal (1994) the NPE-1968 suggested for highest priority to Science-education and research. As the previous paragraphs reflect, the study of environment was to be pursued along with the study of science. Concern for environment is also seen for in the Teacher Education Curriculum Framework published by NCERT in 1978. The curriculum framework suggested following major objective to be Kept in view at all stages of teacher education.

"3-15 Not only use but also help in the conservation of environmental resources and preservation of historical monuments and other cultural heritage." p.13.

Similarly, the National Policy on Education-1986, also highlighted the concern for environment in following words:

"Education and Environment

8.15 There is a paramount need to create a consciousness of the environment. It must permeate all ages and all sections of society, beginning with the child. Environmental consciousness should inform teaching in schools and colleges. This aspect will be integrated in the entire educational process." (p. 23).

One of the important recommendations of Rammurti Committee, 1990, according to Aggrawal (1994) was that, respect for all types of living organisms should form an important part of subject content and process of education. Similarly according to Aggrawal (1994) Janardan Committee, 1992 suggested for encouragement of rural education on the basis of Gandhian thoughts. As we know that Gandhian thoughts clearly reflect the respect for all organisms and also a clearcut concern for environment.

Further, according to Aggrawal (1994), the revised NPE 1986, 1992 also suggested that National Literacy Mission should be evolved according to the national goals like 'environmental conservation' and so on.

Keeping this in view the Curriculum Framework for Quality Teacher Education (1998) has included the issue of environment in the teacher education programme. Under general objectives of pre-service teacher education it has been specifically mentioned that general objectives could include the following:

> "—To sensitize teachers and teacher educators about emerging issues, such as environment, ecology, population, gender equality, legal literacy, etc.;" (p. 25).

Further under curriculum Framework for Teacher Education for secondary stage "Environmental Education" has been put as one of the optional courses (p. 48) and suggestion has been given to include it in the curriculum.

For Non Formal Education it has been mentioned that curriculum for functionaries would be need based and among certain essentials, "— environmental and population related issues;" (p. 94) also needs to be considered. Even for training of adult education functionaries it has been suggested that one of the major objectives may be to; "enable them to understand environmental and population concerns;" (p. 96-97).

Regarding school education, the National Curriculum Framework for School Education Discussion Document published by N.C.E.R.T. (1999) has mentioned the following general objectives of elementary and secondary stage.

"School curriculum should, ..., help to generate and promote among the learners: ... .

6. Understanding of the environment in its totality (natural and social, and their interactive processes), the environmental problems and the ways and means to preserve the environment;" (p. 29).

The framework while raising some issues regarding science and technology at school level mentioned the following about integration:

> "Teaching of science at the primary level is fully integrated with environment (natural and human). Coming to the secondary level, it has been presented as a unified science where there is integration of various science principles, laws. However, this integration as available in the existing science books is only at the superficial level. This integration was done with the purpose of: *(i)* ... .; *(ii)* accommodating important areas of interdisciplinary multi-disciplinary nature such as agriculture, oceanography, cosmology, health and nutrition and environment," (p. 49).

Further in case of social sciences it has been suggested that, "Human-Environment Interactions-Processes and Patterns" (p. 54) should be integrated in the social sciences. This can be introduced in such a manner that at pre-primary stage it should be based on, a variety of activities and programmes directed at providing ample opportunities for the child to interact with other children and with his/her environment. At the primary stage i.e. at class I level social sciences "should be called as "Environmental studies". In Class I and II children should be introduced to the environment as a whole, without making a clear cut distinction between natural and human elements ... In classes II to V children should be introduced to the natural and human aspects of the environment ... children should be given opportunities to investigate the changes in their own lives and those of their families ... school should be given full autonomy at this stage to use locally developed curriculum and locally available resources for teaching of Environmental studies." (p. 55).

For secondary stage it has been suggested that, " ... the processes and patterns of human-environment interaction in India and the issues of environment and development may be included...to provide a broad understanding" (p. 56). At senior secondary stage it has been suggested that under Health and Physical Education, "Improvement of local sanitation, and public health services should form part of field work..." (p. 75). At this stage one of the main objectives should be promotion of problem solving and creative thinking abilities in the citizens of tomorrow.

Conclusion

A review of above sections clearly reveals that Environmental Education (EE) in modern world has a recent origin as it's foundation was laid down in 1970's only. In independent India also it has developed at the same time. But the suggestions given by Kothari Commission show that concern for environment in India is not a new phenomenon. It was part of science education in 1960's and not as a separate subject. It was integrated with science, and social science. Therefore, it can be said that concern for study of environment in independent India is not a new thing. On the other hand the review of above sections also reveal that concern for environment in education was highlighted in National Policy on Education-1986. Thereafter, attempts are being made in all sectors of education and especially school education (pre-primary, primary and secondary) and teacher education to promote environmental awareness among students as well as teachers. Hence, attempts are being made to develop a environmentally conscious society, so that the deteriorating natural environment can be protected and a pollution free environment can be developed in near future.

REFERENCES

1. N.C.E.R.T. (1970) '*Education and National Development*' Report of the Education Commission 1964-66, Ministry of Education 1966, Govt. of India, New Delhi.
2. N.C.E.R.T. (1978), *Teacher Education Curriculum a Framework*, New Delhi.

3. M.H.R.D. (1986). *National Policy on Education - 1986,* Department of Education, Govt. of India, New Delhi, May 1986.

4. Sharma, R. C. Mahajan, B., Premi, K.K., Nuna, S.C. & Menon P. edited (1994). *Source Bool on Environmental Education For Elementary Teaching Educators, NIEPA* (with the support of UNESCO-UNEP-IEEP), New Delhi, p. 278.

5. Sharma, R.C. (1994), 'Environmental Education Asian Perspective' in Sharma, R.C., Mahajan, B., Premi, K.K., Nuna, S.C., Menon, P. edited (1994) *Source Book on Environmental Education For Elementary Teacher Educators,* NIEPA, New Delhi.

6. Aggrawal, J.C. (1994) "Rashtriya Shiksha Niti, (1994 Tak Hue Sudharo, Sansodhano Va Parivartano Sahit), Prabhat Prakashan, Delhi pp. 17, 20, 68, 77 & 87.

7. Khosla, D.N. (1998) *Curriculum Framework for Quality Teacher Education,* NCTE, New Delhi.

8. N.C.E.R.T. (1999) *National Curriculum Framework for School Education—A Discussion Document,* New Delhi.

9. D. D. Metro (2001) '*Earth Matters*', 6th March 2001.

10. D. D. 1 (2002) '*National News,* 8.30 P.M., 20th Marth 2002.

Socio-Political and Administrative Dimension

7

Environmentalism

A Global Concern

Prof. Priyankar Upadhyaya*

The ideational lineage of environmentalism (the concern for environment protection) may be traced back to various ancient thoughts particularly in Indian tradition. But in practice the long sweep of civilization has been witness to a relentless human interference with the non-human world leading to irreparable damage to the natural planetary world and its life-support systems. The unbridled environmental corrosion had assumed alarming proportions bong back; its severity has dawned on the humans only in the last quarter of the 20th century provoking fears that an era of global instability looms on the horizon.

Since early seventies one could discern a gadual but decisive transformation in favour of eco-friendly conduct of human affairs. This has been a departure from hitherto dominant mindset defining progress as the human triumph over nature. This could be a welcome paradigm shift. Transcending national borders, the environment concerns have imbued the entire range of concepts relating to human progress and development. Sustainable development embodying environmentalism has been the latest buzzword or mantra. Be it official discourse of NGO activity environment has been the watchword. Even the Brettonwoods

* **Professor of Political Science, Co-ordinator, Malviya Centre For Peace Research, Banaras Hindu University, Varanasi.**

institutions have started referring to environment as the touchstone for offering monetary assistance. (World Bank Report: 2000)

The critical question is whether the much-hyped environmental concerns are likely to bring any tangible difference given the present frame of capitalist development and globalised commerce. And even though the raised consiciousness about environment would certainly make a difference in the long term but by then the damage would be beyond rectitude. How to avert this man made disaster ? There are no easy answers nor it is possible here even to sketch the entire magnitude of the problematic. Given the limited scope of this presentation it would however be instructive to overview the environment problematic as it has transpired globally and what is imperative is case of India.

Convergence and Divergence

It is possible to divide the contemporary environmental problems in two broad groupings. While problems like global warming, ozone hole, and acid rain are essentially the creation of the rich industrialized North, the ones such as water, air and sanitary pollution are being attributed to the poor South. The North/ South divide also has significant polemical dimensions. For instance Environmentalism is projected as a product of 'postmaterialist' values among the upper middle class population of North America and Western Europe reflecting a shift from physical sustenance concern to qualitative improvement of life and also an assertion of self against the systemic norms. So much so that Anna Bramwell commented wryly in her that "Only the maligned Western would has the money and the will to conserve its environment. It is the 'Northern white Empire's last burden, and may be its last crusade". (The Fading of the Greens.) But irrespective of who is more accountable for the present state of affairs the truth is that the environmental degradation has reached a point wherein human survival warrants urgently needs concerted resolve and action both public and private from all quarters.

Globalised Commerce and Environment

Another contention central to environmentalism is whether the contemporary form of globalization holds any good to the quest of sustainable development. On the one hand it is argued by the proponents of globalization that without adequate economic development, resources would not be available for much needed investment in critical area of environmental protection. "The key to gain sustainability is not to produce less but to produce differently" and the best policy of environmental protection will help and not just economic development: (World Bank: 1992). Seeking global effort to combat climate change, the United States, of late, is projecting environment as central to its national interest as well as for the entire civilization. Here to the argument is in favour of globalization that the developing countries need a greater competitiveness and access to new technologies which would lead them to greater prosperity without sacrificing air, drinkable water and livable cities.

Conversely there are opinions discounting the contemporary globalization as a asymmetrical development wherein the developed world (or the First world) is gaining at the cost of the Two-Thirds World (a graphic way of describing the Third-World—as the 2/3 of world population resides in the Third world). Relentless drive for industrialization combined with menacing pace of urbanization has already created insurmountable problems of sewage disposal, environmental sanitation, traffic congestion, pollution and which warrant greater resource sharing between the developed and the developing world. In the Ministerial conference of the WTO in Seattle last November, thousands of NGO activists protested against what they see as the Two's environmental blindness revealing instance is the steady rise of the value of global trade in forest products over the last few decades, climbing from $ 47 billion in 1970 to $ 139 billion in 1998. The environmental impact of freeing trade in forest products is another controversial issue looming on the trade horizon. Similarly the globalised market drive encourage over fishing for export market which entails depriving small-scale fishers of their bounty and also raising the fish price level beyond the means of local people, placing a number of commercially valuable species at growing risk of over-exploitation and even extinction. The wildlife trade is valued at $ 10-20 billion annually; at least a quarter of which is thought to be illegal. The list includes, on an annual basis: 40,000 monkeys and other primates, 2-5 million live birds, 3 million live farmed turtles, 2-3 million other life 3 reptiles, 10-15 million raw reptile skins, 500-600 million ornamental fish, 1,000-2,000 raw tons of corals, 7-8 million cacti, and 9-10 million orchids. More than 30 new infectious diseases have emerged over the past 20 years, including AIDS, Ebola, and hepatitis C and E. According to the World Health Organization, 'environment changes have contributed in one way or another to the appearance of most if not all' of these diseases. (World Watch Report: 2000)

The unattended incompatibilities between ecological imperatives of local communities and the pro-corporate economic practices directed by the Brettonwoods institutions (World Bank & IMF) create a paradoxical situation. And notwithstanding the recent reforms aiming to make the world trading system environmentally sound (World Bank Report: 2000) there is much to be done to rectify the environmentally harmful pay outs specially in the area of fisheries, agriculture, energy and forestry subsidies.

Philosopher Rudolf Bahro emphasizes on changes in existing patterns of production, consumption and distribution as a radical solution to the ecological crises. According to him the ecology crisis is insoluble under capitalism. The need thus is to reform the economic institutions plus reinforce the International Environmental Institutions (UNEP) much in the pattern of WTO.

All this negative view of globalization should not divert us from the act that the greater connectivity has offered us a historically unique confluence of shared glogal environmental concerns transcending national and regional borders and uniting I/o's and NGO's on common concerns such as depletion of natural resources and increasing world population, deforestation, soil erosion, global warming and climatic change due to unbridled emission of greenhouse gases, thinning ozone layer, pollution etc. Earth Summit and its follow up at Berlin and Kyoto is a case in point.

Indian Situation

Much like other developing countries, in India too the flawed development policies have compounded the worsening eco-system. Be it unplanned growth of slums, inadequate public transport systems, haphazard growth of industries, encroachments, improper sewage systems or lack of public awareness on garbage disposal,

— all have led to a conspicuous environmental damage in the past few decades. And much in line with the western capitalist countries the industrialization and the resultant developments in transportation, refrigeration, air-conditioning, too has contributed to the disruption of ecological balance. There is an increasing threat of exposure to carcinogens and by zenobiotics; the talk of higher incidence of heart diseases, cancer, and other evils. So notwithstanding the growing life expectancy, the quality of life of those living has been lowering down with the unbridled growth of industries. So much so that one wonders whether the humans were better off without all the technological and industrial development.

For people, developing cities across the country, have found no respite from dirty water, smoky skies, polluted air, degraded lands and congestion. The food we eat, the water we drink and the air we breathe are all highly contaminated. Several of our rivers are dying; cities choking and underground water tables rapidly depleting. In over 20 water basins, people reportedly face water shortage. Out of the total forest cover of 0, 76 million sq. km the country has lost some 5,500 sq. km during the past two years alone. Thanks to indiscriminate spraying of some 50,000 tones of pesticides annually for the past 30 years in order to improve agricultural yield, our fields have been poisoned to the extent that even if spraying is stopped now, the effect will linger for another 50 years, The adverse impact in terms of ecological hazards is estimated at close to 10 per cent of the GDP. Parks and open greens, the only lung spaces in our cities, are recklessly being encroached. The migratory birds that visit our lakes are increasingly being poached. The gradually diminishing chirping of the birds reminds of the famous book of Rachael Carlson "Silent Spring" which had warned the West of the demeaning aspects of unchecked environment decay. And one feels that if nothing drastic was done in India too there is a threat of spring arriving and going in silence.

According to the Central Pollution Control Board (CPCB) 70 per cent of the air pollution is caused by emissions of two-stroke engines. A new study on pollution conducted in Calcutta reveals that, besides facing all other evils of air pollution, traffic police chronically exposed to harmful automobile exhausts also run the risk of alterations in the basic structure of their genes. Only a little more than half of the 2.25 million tones of solid/medical wastes generated daily gets cleared, although not systematically disposed of. Nearly Rs. 20,000 crore gets frittered away annually to treat environmentally acquired diseases. The abysmal state of affairs prevailing in several parts of the country came in sharp relief at a recent conference of state environment and forest ministers in Delhi. Notably, in Bihar, sewage treatment plants do not function while in Delhi, they are under-utilized and hundreds of unauthorized colonies have no sever system. Limited disposal facility for solid wastes arising in domestic and other industries is another pressing environmental problem, causing extensive pollution and threatening human health, Seven-tenths of the population have no access to sanitation. Just 71 of the 212 Class One cities have water collection systems to take care of 12,000 million liters of municipal wastes generated daily. The quickening pace towards environmental self-annihilation has been responsible for destroying rain forests and the green cover.

To clean up to polluted rivers in the country, River Action Plans started in phases but they still remain incomplete. On public interest litigation filed, seeking monitoring of implementation of the Supreme Court's direction to maintain cleanliness, the Delhi High Court recently fined the Delhi government for allowing the Yamuna to get polluted. It has also issued notices to civic bodies to intensify their drive against irresponsible garbage dumping by residents, hospitals and other institutions. The precarious garbage situation in Delhi prompted the Supreme Court to direct all major hospitals to incinerate their waste. Five years later, as reports indicate, hardly any of them have bothered to even install incinerators. All garbage of some nursing homes and government hospitals in many cities are just dumped into nearby areas. Not only are syringes, glucose and blood transfusion pipes recycled but also is cotton picked up from hospitals and nursing homes use to make mattresses.

The Way Out

Environmental awakening in India got under way in the early 1970s as an open campaign to protect the trees in the Himalayan region with the Chipko Movement, spearheaded by Sunderlal

Bahuguna and others. Other crusades-related Tehri/Narmada dam projects, Chilka lake etc. — have all received wide attention. There are success stories many of them involving women asserting their rights to clean environment. But the big question has been how to integrate the environment considerations with developmental aspirations which can lead to sustainable development. Since environmental protection is directly linked to economic progress, concerted efforts hold the key to a cleaner future. But forging an environmentally sustainable society is about more than economics, and far-sighted economics is about more than reducing restrictions on the movement of goods and money'. Creating a global society fit for the 21st century will thus require not only reform of economic institution, but also a strengthening of international environmental institutions so that they can act as an ecological counterweight to today's growing economic powerhouses.' As a solution to the global environment governance, one idea gaining ground is to upgrade the UN Environment Programme into a World Environment Organisation (WEO), on par with the WTO.

REFERENCES

1. Weting H., (1984) (ed.): *Environmental Warfare; A Technical, Legal and Policy Appraised,* Philadelphia, Taylor and Francis.
2. David Held and Others (1999): *Catastrophe in the Making Globalization and the Environment,* California, Stanford University Press.
3. Paul Brown (1998): *Warming: Can Civilization Survive ?, Hyderabad, Universities Press* (India) Limited.
4. Ashok Swain (1996): *Environmental Migration and Conflict Dynamics; Focus on Development Regions,* Third World Quarterly, Vol. 17, pp 959-973.
5. Gal Omvedt (1994): *Green Earth; Women Power, Human Liberation,* in Shiva Vandana (ed.), Close to Home; Women Reconnect Ecology, Health and Development Worldwide, Philadelphia, PA, New Society Publishers.
6. Shiva Vandana (1982): *Staying Alive; Women, Ecology and Survival in India,* Kali for Women.
7. Fredmann John, Haripriya Rangan (1993): *In Defense of Livelihood; Comparative Studies on Environmental Action, Kumarian Press.*
8. *World Watch Report: 2000.*

8

Environmental Issues

A Social Scientist's Concern

Dr. Rachana Srivastava*

"One can not be truly human and civilized unless one looks upon not only fellowmen but all creation with the eyes of a friend .... It is said in country after country, progress should become synonymous with an assault on nature. We, who are part of nature and dependent on her for every need, speak constantly about "exploiting nature" ...". I remember Edward Thompson, a British writer and a good friend of India, once telling Mahatma Gandhi that wild life was fast disappearing. Remarked the Mahatma: "It is decreasing in the jungles but it is increasing in the towns."

The passage cited above is excerpted from the famous speech delivered by the then Prime Minister Mrs. Gandhi while addressing the plenary session of the first UN Conference of Environment held in Stockholm in June 1972. Things have not changed a wee bit and man still remains the most dangerous species in the world threatening not only wild life world over, but also all other forms of life. The sad truth is that he still nurses the illusion that he can survive even if the quintessential life support system that sustains him is destroyed.

A belated realization that man must learn to coexist with the animal and plant kingdom and litho, hydro and troposphere has

* **Reader & Head, Department of Political Science, Vasant Kanay Mahavidyalaya, Kamachha, Varanasi.**

spurred India and other countries of the world to take several measures to preserve the ecological balance and save the fast dwindling varieties of wild life from the brink of extinction and protect environment from excessive pollution.

The primitive man however lived in harmony with nature. It is only after the spread of civilization and the progressive exploitation of nature that the ecological balance was threatened. Hunting animals in a big way, alteration of the environment, destruction of the habitat, pollution of the land, air and water, the human population explosion—all these have led to a number of environmental problems. Pesticides and industrial effluents have contaminated the aquatic eco-system, a vital link in the survival of the animal species. It is said that since the 17th century about 120 species of mammals and 150 of birds have become extinct in India only.

With a land mass of 329 million hectares and a coastline of 7516 kms, with oceans lakes, rivers the mighty Himalayas and other mountain ranges, the deserts of Rajasthan, the plateaus, the wetlands and the islands of Andaman and Nicobar and Lakshadweep, India is home to an amazing variety of fauna and flora. There are about 75,000 species of animals, of which 340 are mammals, 1200 are birds, 420 reptiles, 140 amphibians, 2000 fishes, 50,000 insects 4000 mollusks and several other species of vegetables. At present 81 species of mammals, 38 of birds and 18 of amphibians and reptiles are considered to be endangered in India.[1]

Laws have been made both at national and international level to conserve the environment but laws alone are not sufficient. Awareness in all sections of community has to be created. People must realize that they have a stake in preserving the balance of nature. Luckily we have a vigilant media, vocal environmental groups, NGOs and other individuals like Medha Patekar and Sundar Lal Bahuguna who are creating public awareness regarding environmental issues.

We must ponder sincerely upon the observation made by Romain Grey in his thought provoking introduction to Time-Life book, "Vanishing Species"— "There is nothing in nature to prove that it cares more for one human species than for our daffodils. We may one day vanish as quickly and as radically as thousands of other breeds before us ...Mother Nature has no mama's darlings... When the balance of nature is threatened, it always finds a way to restore that balance, at whatever cost. If endangered by us, nature will strike back...."

It is in this context that there appears to be ever more the need for making Enviornmental Study a branch of Social Science. Social Science deals with various sybsystems that are working in and consequently having an influence on this vast social system. Ecology or environment is an important sybsystem of this broad social system which certainly has a great influence on human life.

What is Environment?

All that surrounds man and his family of living species and all that governs and influences life come under the broad definition of environment. This includes atmosphere, lithosphere, hydrosphere and biosphere. At the beginning of our civilization man lived in perfect harmony with his environment. However as man progressed he tried to be the master of environment and in this process we destroyed our inheritance resulting in earthquakes, acid rains, landslides, global warming and man made tragedies etc. Infact whole of civilizations got erased in some instances.[2] It has become a painful but frequent phenomenon today that monsoons, cyclones, tropical storms, torrential rains hit coastal and low lying areas killing hundreds of residents, rendering homeless thousands of people, flooding large areas, destroying cattle, submerging crops, cutting off communications, disrupting the lines, cutting off roads and traffic, inundating the reservoirs, breaching dams and tanks, marooning the villages, devastating houses etc.

What are the Challenges?

A World Bank report released in 1998 says that more than 40,000 people die prematurely every year in India alone because of Air Pollution. Patients with respiratory diseases and allergies have doubled since 1990. Auto exhausts and industrial emissions have dangerously crossed the limits. 21 century will be less in oxygen and more in carbon-di-oxide and other pollutants. Dust and particles of carbon coated with toxic gases have exceeded the standard of safe levels. Aspirational Bronchopneumonia has become common due to pollutants from petrol cars and vehicles which throw out hydro carbons, oxides, nitrogen, black smoke etc. Lead in petrol affects central nervous system, carbon-di-oxide reduces the ability of blood to carry oxygen and provokes heart diseases, suspended particles of dust and carbon cause respiratory diseases, Aromatic hydrocarbons from diesel engines cause drowsiness, eye irritation, cough etc., sulphur dioxide causes bronchial problems, Benzes causes lung cancer, Nitrogen oxides cause colds, coughs , bronchial irritation etc.[3]

There was a time when we worshipped Mother Nature. Indian scriptures are replete with incidents where people are directed to consider forests, trees, rivers and mountains as sacrosanct. Cutting of trees was prohibited, although on religious grounds. May be because religion could appeal more to people's minds than science. All this averted any conflict from ensuing between man and his environment. There was a harmonious balance between man and nature. Nature nourished her children and the children worshipped their mother. Gradually this Mother-child relationship was replaced by master-slave relationship. Even during British rule, the makers of modern India the likes of Swami Vivekanand, Dayanand Saraswati, Ramkrishna Paramhansa, Mahatma Gandhi and many others tried to bridge the gap between man and environment. The idea of 'Sujalam Sufalam Shasya Shymala Mataram, Vande Mataram' was mooted and propogated. But man's basic desire to enslave every thing whether animals or difficult physical terrains soon overcame this reverence. In his quest to make life more and more luxurious for himself, man destroyed the ecological balance. Forests were cut, rivers were diverted, mighty mountains were conquered, factories emitting harmful gases were setup, oceans were polluted and even space was not spared, motor cars and many machines which were chief pollutants of atmosphere were invented and finally nuclear weapons that could destroy the whole earth not once but a number of times were built and piled up. In his bid to master environment man eventually destroyed it. Just as a human body, cannot function properly if it's various organs and limbs lose coordination, similarly the environment gets affected if ecological balance is disturbed. It is evident from floods, famines, landslides that are some of the results of the unmindful cutting of forests and trees. Fish and turtles are dying due to chemical pollution of oceans and rivers. Another example of conflict between nature and man is the spread of mad cow disease resulting from man's greed to earn more by getting more milk from dairy animals through harmful milk-inducing injections. Man himself is an important part of biosphere but sex determination test resulting in female foeticide particularly in developing societies like India is one more incident of environmental challenge. And now we have human cloning, which if used for medical and remedial purposes, may be justifies, otherwise this too may prove to be bane to mankind. In present scenario when environmental challenges themselves are a threat to man, the whole issue of environment has been hijacked by the developed world to exploit the developing world and to impose their terms and conditions on them. WTO meet at Seattle, USA and the agitations there by environmental agencies against the developing countries is an evidence of this new trend. Environment protection has always been a global concern but now the issue has been politicized. Developing countries have been accused of floundering environmental guidelines in producing goods. Although this is true to quiet an extent, yet developing countries cannot be considered responsible for whatever harm has already been done to the environment by the developed world. And secondly developing countries lack the required technology and basic infrastructure needed for environmental protection. This situation requires mutual cooperation among developing and developed countries and not conflict, as is the case. The interest of the people of developing countries goes beyond merely dealing with environmental issues—from climatic change to ozone depletion and soil erosion etc. (litany of human created destructions during the past 300 years in the name of progress and development), to more fundamental issues, which are at the root of all earthly problems. These are the problems of development: growing population, poverty, technology transfer, unfavourable trade systems, the third world debt of $ 1.3 trillions, lifestyles of the people of northern hemisphere, piling of armaments (at an annual cost of $ 1.4 trillions) and its impact on environment and development, archaic geographical boundaries, and a host of social problems.

What are the Causes?

The social and political unconcern on the part of the people in general and governments in particular has been mostly responsible for environmental degradation. Due to this deliberate ignorance the environment bomb seems to be on the verge of exploding. Social and political inaction has led to biological triggering of this bomb. Chief factors responsible for the creation of such an alarming situation as this are: the green house effect and global warming resulting in inundating of vast regions and even whole islands by see water, on the one hand and droughts and famines on the other; ozone depletion due to the use of CFC in A.C. and refrigerators etc.: the world wide phenomenon of deforestation which is causing soil erosion, droughts and lowering of underground water-table; an unprecedented level of industrial pollution which is degrading air with cancer causing toxic elements and causing acid rains and smog etc.: water pollution due to chemical effluents and sewage waste; land degradation resulting from indiscriminate use of chemical fertilizers and deadly pesticides; and finally the potential radio active fallout from the increasing number of nuclear power plants.[4]

Although from the beginning of our civilization, mankind has been polluting the environment, it was able to fight back and regenerate itself. However, today, as we equate development with industrialization and the production of material wealth and as the pressure of population is increasing, our environment has started to loose it's ability to regenerate itself. Our industrial wastes are polluting the atmosphere with oxides of carbon, sulphur and nitrogen and suspended particles etc. Effluents and solid wastes of chemical industries are affecting land, ground water and river water. Radioactive wastes dumped around carelessly are endangering human species. Untreated urban sewage is affecting our precious rivers, lakes and other water bodies since it contains phosphates and nitrates. This will give rise to growth of algae and melds in tanks and lakes, leading to entrophic conditions.

A properly conceived and executed planning can play an important role in reducing industrial and urban pollution. However, because of vested selfish interests, ignorance on the part of the people, lack of expertise on the part of planners, bureaucratic hassles and rampant corruption, we are unable to have a proper town planning. Not enough space is provided for parks and roads, industries are located sometimes within the residential area and no green belt is provided even when it is easy and economical. Increasing population especially in some parts of the world like India is also a cause of environmental disbalance. Today we have 6.2 billion people in the world claiming their share of fast depleting natural resources, food water and even ozone 70 per cent of these people live in the developing countries in Asia, Africa and Latin America, mostly, in subhuman conditions and have only 30 per cent of the world resources while the 30 per cent of world population has remaining 70 per cent of world resources at its disposal.

What We Must Do?

We should realize that development without environment is not possible. We cannot have one at the expense of the other. Sustainable development should be our new mantra[5]. The impact of environment of socio-economic dimensions has increased and the problem is how to control it. No doubt, man himself is the enemy of his environment, but can the society service if this is allowed to go unabated! The hope lies in adopting determined measures urgently. As the matter involves the society, it becomes the duty of the social scientists to make a critical assessment of environmental damage due to developmental activities. Priorities of development schemes should be carefully decided by rational means, keeping long-term perspective in view, rather than by short-term political expediency. Nowadays we find judiciary performing the duty of a social sentinel preventing governments from going ahead with such developmental plans as may prove to be more devastating than beneficial Supreme Court's strictures against Sardar Sarovar Project and Tehri Dam Project are examples of this. Ministry of Environment, Government of India was instructed to furnish Rehabilitation plans for those who would be rendered homeless due to the building of dams. Afforestation plans were also required to be chalked out to compensate for the submergence of vast forest areas. Supreme Court's decision regarding the use of CNG in vehicles in Delhi, considered to be third most polluted cities of the world, ban on industries around Tajmahal in Agra etc. are some of the examples of the important role that judiciary has been playing in conserving environmental balance in India. Judiciary had to act because of the inaction on the part of the administration. The insensitivity on the part of the policy makers and bureaucrats is a serious matter. There is need to develop problem solving approaches that integrate principles and methodologies of ecology, economics and social science. Economic costs of required environmental measures need to be quantified and industry should support environmental research. There is need to sensitize people towards environmental issues. we save money and jewellery to pass on to our next generations in order to make their lives comfortable but have we ever given a thought to as to what kind of environment we will be passing over as legacy to our future generations. It is ultimately the younger generation that will have to face the impact of environmental degradation and so they must learn to care for the same from the very beginning. This can be done through both formal and informal education. Through our various festivals like Holi, Diwali, Baisakhi, Sankranti, Onam, Bihu, Pongal, Vasant Panchami, Navaratri, Dev Dipawali etc. we inculcate in them reverence for various forces of nature. However the extent of damage to environment necessitates formal education to children regarding ecological balance and environmental conservation. Only when they have a basic understanding and care of the natural environment will they realize the meaning of sustainable development. The conflict between man and nature can be resolved only through people's voluntary participation in environmental management. No government can cope with the task of environmental repair alone. Without educating people environmental laws (there are over 400 hundreds of them), cannot be enforced.

The gravity of the problem and the need to search for possible solutions to it requires the development of the study of Environment Science as a separate branch of Social Science. As social scientists we have a responsibility towards the conservation of local and global environment as well as social environment.

Third World Concerns and Environmental Issues

Three-fourths of the globe is occupied by the developing countries and so they should realize their role in environment degradation and conservation. The people of developing countries only seek means of survival—water, food, renewable sources of energy, education and mans of livelihood. Their only wish is to live in harmony with the land, trees, birds and animals as they have done since human beings evolved as homosapiens, four millions years ago. They hope to create their own future, instead of leaving it to destiny. It will be created only by those who will provide every child a livable, sustainable and enjoyable future so that evolution progresses without violence and bloodshed. For this the subject of Earth protection and Environment safety is not only of paramount importance but also topical. Yet the developing countries would be seriously jeopardizing their future development prospects for years to come if they bind themselves to international agreements on issues related to global warming and biodiversity conservation without ensuring major commitments from the industrialized countries on reduction of carbon-di-oxide emissions and fixation of equal per capita permissible emissions on a global basis.

If, for instance, the developing countries were to agree to forest conservation under biodiversity convention, they would be conserving a major carbon reservoir, which the climate convention also seeks to protect. But are the industrialized countries then prepared to reduce their carbon-di-oxide emission ?

There must be a link between bio-resources and biotechnology. Some countries would like to see tropical forests declared a global asset, as there is an urgent need to preserve them as sinks for carbon-di-oxide emissions. But how can forests, which are the resources of individual sovereign countries, become global resources while technologies, which can help reduce those emissions, remain in private hands?

The main responsibility for the state of environment today lies with the developed world, which must take corrective action to change their wasteful consumption patterns. The developed countries are stressing the responsibility of developing countries in the resolution of environmental problems of global significance and are also trying to shift the burden of protection and regeneration of the environment to the developing countries. This gives rise to a new perspective of environmental problem, Can we delink poverty and development from environment issues? Should environmental standards be harmonized at the global level? And same standards be applied to developed and developing countries? Should we adopt western viewpoint to tackle the problem of developing countries, which are unique to this region ? What may be an environmental issue to the developed world may be a human right issue or question of right to livelihood and right to food for the people of developing world. No one disputes that a pollution free environment is as necessary for survival in the developing part as it is in the developed part of the world. However it is inequitable and impractical to ask the developing countries to enforce the same rigid environmental standards as the developed nations, as was recognized in a study by GATT on the subject of trade and environment. The study had concluded that interference is seldom, if ever, the best way of achieving a particular domestic environmental objective. In a similar vein the same was recognized in Principle 11 of the 1992 Rio Declaration on Environment and Development, which says, "Standards applied by some countries may be inappropriate and of unwarranted economic and social cost to other countries, in particular developing countries." It is also agued that environmental insistence is a neat way of trade advancement. It is clear from the fact that the Japanese Government plans to limit "tied ODA loans" to projects related to anti-pollution and other environmental projects and measures.[6] In fact, environment related manufacturing industry is the fastest growing in the developed world. The Human Rights activists from the third world have argued that it is a welcome sign that environmental awareness is on the increase, but there is the other angle which calls for caution, namely, to be alert to see that the slogan of globalization and the environmental lobby do not stifle the industrialization and growth of economy in the developing countries. If in the name of environment conservation small industries would be closed, it would not only give rise to unemployment but would also throw open the small industries sector to the big and multinational sector on the fallacious ground that the latter alone have the resources to provide the anti-pollution mechanism, when the highest pollutants are the units in the big sector, It is they that have polluted Taj and rivers like Ganga and Yamuna and the drinking water reservoir. Small industries sector contributes about 40 per cent of GNP and livelihood of 177 lakh people (in India) depends on it.[7]

These are some of the issues related to environmental problem. Shouldn't third world perspective be taken into consideration ? if we sincerely wish to tackle the environmental issue and diffuse the environmental bomb, we must study all the dimensions of the problem.

A Social Scientist's Concerns

Environmental issues are issues not only of rising levels of carbon and sulphur in the air but they are also issues of sustainable development, human rights, sensible policy making and creation of public awareness. As such they are just as much matters of concern for an economist or a political scientist or a sociologist or a historian or a human right activist; just as they are for an environmentalist or a geo-scientist.

Social Scientists are concerned with social environment, with input (demands and support) and output (decisions and policies) and also with physical environment. Man cannot function in vacuum. He works in coordination with his social and physical environment; both at inter state and intra state level Continuous socialization with environment is required to maintain compatibility. We have already discussed that economy, development, poverty and social policy making cannot be delinked from environmental issues. Agents of social and political change, i.e. Legislature, Executive, Judiciary and Bureaucracy need to be sensitized both on environmental issues and social engineering in a system. Every system has its own peculiar conditions and policy makers and implementing agencies should keep balance between development and environment as per the demands of their system. Only then will there be a proper functioning of stress free system. Just as social environment may be riddled with stress if demands arising in it are not paid attention to by the government, similarly physical environment too is full of stress which may eventually lead to a breakdown if the balance is not maintained.

We have to educate common man about ecological balance. Though in all the schools environmental protection is taught, our students are not getting the message. We need to take urgent steps to remedy this. There are many questions to be answered, many theories to be developed, many perspectives to be investigated and various dimensions of environmental issue to be examined and analyzed. Social Science is the study of society. If society perishes we too perish. The major task before us is to rejuvenate the society and play our role in diffusing the environmental bomb.

REFERENCES

1. Competition Success Review, September 2001, p. 17.
2. Report in India Today, February 11, 2002.
3. Rao. N.P. 'Global Strategies of Clean Environment, Safe Earth, Disaster Management, Sustainable Development and Quality Life (National and International Obligations and Priorities) Delhi, 1998, p. IV.
4. Shankar, Jagan—Social Problems and Welfare in India. New Delhi, 1992. p. 331.
5 According to the World Commission of Environment and Development, 'Sustainable Development is That Development That Meets the Need of the Present Without Compromising the Ability of Future Generations to Meet Their Own Needs' Mainstream, December 23, 2000. p. 29.
6. Ibid.
7. Ibid,p.30.
8. Yojana, New Delhi, Various issues, May, July, October and December 2001.
9. CSR, November, 2001, p. 331.

9

Environmental Issues
The Political Dynamics

Dr. Bhavana Trivedi*

Mankind is a part of nature and life depends on the uninterrupted functioning of natural systems which ensure the supply of energy and nutrients ...civilization is rooted in nature which has shaped human culture and influenced all artistic and scientific achievement, and living in harmony with nature gives man the best opportunities for the development of his creativity, and for the rest and recreation.

–The World Charter for Nature

(Adopted by the General Assembly of the United Nations on October 29, 1982)

The serious threat to human survival today undoubtedly is environmental degradation world wide. Environmental problems can not be studied/understood as an isolated problem for the mankind because it leads to severe social and economic consequences by degrading the human health and the resources as a result of the so called development. It has also posed challenges to the policymakers, social activists, academicians, environmentalists, human rights experts and business community as well.

Post World-War scenario witnessed a sharp deterioration in human natural environment i.e. air and water pollution, noise

* **Senior Lecturer in Political Science, Arya Mahila Degree College, Varanasi.**

pollution, accumulation of all kinds of waste products and discards over wide areas, extensive use of many natural components and extinction of some wild life and plant varieties, degradation of soil and other forms of disruption of natural balance such as climate change, stratospheric ozone depletion, biodiversity loss and global warming. Keeping in view this deterioration, in the 60's and 70's efforts were initiated to protect and conserve the human environment. But the situation gradually became grim with the dawn of globalization and with the upsurge of market oriented economy which added new challenges to environment. It is ironical that till this date, there is no unanimously resolved policy framework and action programme to save the environment of the world.

Environmental problems noticeably are no longer restricted to the geographical boundaries of nations. The few glaring examples of cross boundary environmental hazards may be cited as the emission of sulfur dioxide from England causing acid rains in Scandinavian countries, soil erosion and flood in India and Bangladesh may be seen as the repercussions of the deforestation in Nepal that is an evidence showing how deforestation in one country reduces the biological richness of the other part of the planet. Similarly chemicals and gases produced on one continent can deplete stratospheric ozone causing skin cancer on the other part and carbon dioxide emissions in any part of the earth be it America or Europe, Asia or Africa hasten global warming and climate change in the whole globe. In such situation of global pollution how the economic and political policies and decisions for development in one part of the world can remain exclusively a matter of concern for one nation only.

There is an intimate relationship between economics, politics and environment. We are aware, that all developmental activities of a nation are designed, framed and governed by economic and political policies of the government. The developmental activities like establishment of industries, dams and irrigation projects, manufacturing of consumer goods, farm and forestry development etc have some or the other environmental implications. The Politics has directly and indirectly over-powered economics and both in turn neglected the environment, with its care, protection and management receiving a third priority. The politics could be seen from local to national and to global at every step of the political system. Most of the times the decisions for the approval of major projects are by and large taken instead of on rational, pragmatic environmental grounds, as is ostensibly claimed, leaving it to people to pay the price in the form of ecological imbalance.

In the present article an effort has been made to study and analyze the political dynamics of environmental challenges and issues in international and national perspectives. The international perspective of the political dynamics related to environmental issues may be summarized as under : Lack of Universally recognized policy framework, Eco-Politics or Eco-Imperialism, Waste Dumping in the Third World by developed nations, Damages caused by political and military conflicts, , Controversy related to Biodiversity and the Matter of Environmental security etc.

There is a lack of commonly shared global strategy on the ecological management. The hidden truth behind this lacuna is the feeling of distrust and non cooperation among nations. The rubric of environment consists in common natural resources shared by all sovereign entities of the world .Thus its protection is the moral responsibility of all collectively but the question arises who will be in the vanguard to initiate the protective steps. Almost none is proactive in undertaking unilateral efforts without being aware of the neighboring nations' intensions and willingness in the matter of multilateral safeguarding of the earth and its resources. Many studies observe that most of the countries are not ready to cut short and minimize their pollution generating practices be it technological or industrial as it would hamper their developmental process and would cost them heavily. They believe that they benefit strategically by encouraging their neighbors to introduce protective measures, consequently yielding a comparative advantage for their own industries. Sometimes due to the different sets of ideologies the leaders of other countries are also calculating in this manner, on this suspicion policymakers are often reluctant to undertake environmental measures to protect the international environment unless they are absolutely sure that their neighbors will undertake the same set of measures. Consequently no unanimously decided green-agenda has been prepared. The US decision to back out of the crucial global warming Kyoto Treaty (1997) is a brutal blow to one of the planets' most important struggle for survival and one such example of global non-cooperation.

The politics as well as economics involved in dealing with the environmental issues at international and national levels is major obstacle to the protection of the global environment. One version flowing out from this is the matter of Eco-imperialism which includes technology transfer from North nations to South nations. The northern nations of the world are economically and technologically well off relatively. They desire to have an upper hand over the less-technologically-developed nations of the southern region by using their technological advancement as a means. In this process the advanced nations of the world make efforts to use their technology as a 'political weapon' for making several political bargaining with the less-developed nations. They use it as Intellectual Property Right and argue that technological development should be treated as commercial activity and those countries who want to utilize it, must pay for it. It is an example of flagrant 'Eco-Imperialism' that the North in such way is enticing the South nations diplomatically into its so called progressive trap. The two-third of the world, already under the crushing pressures of debt and conditions imposed by World Bank, IMF, and their own growing population and poverty, have an undercurrent of fear that the matter of environment may be instrumental in the enforcement of eco-imperialism by the developed nations who will command them to follow the social and economic policies of the former.

There is also a rift between the North and the South over the subjects like 'environmental protection and regeneration funds' that it should not be treated as 'aid' but are more like 'taxes' on over-consumption of the rich nations and the environmental mesh created by them. The contention of developing nation is that the north nations have used the life-sustaining human environment for maintaining their luxurious life-styles by allocating greater share of the natural resources which is a 'common property' of mankind but the north is not readily accepting this on the grounds that the south is also involved. Overall the whole issue has generated ongoing reactions and counter reactions without reaching to any definitive point.

Similarly the dumping of waste in the Third World by developed nations is also causing ecological challenges to these countries. The rich and developed nations of the North have a tendency to dump their hazardous wastes in the name of so called 'imported items' into the poor developing nations. They transfer obsolete technologies, old medicines, and agricultural fertilizers to these countries. The developing countries of the south having little expertise accept technological support and financial assistance from the North to fight against poverty and underdevelopment. Political scientists have argued that inequality can affect the environmental security of the globe in general when the poor will overuse and degrade local environmental resources in ways that inflict negative effects on everyone, rich or poor. Many marginalized people, struggling with day-to-day survival, use the few resources they have inefficiently and unsustainably which results in increased deforestation, soil erosion, and depletion of farm land adding to global environmental pressures. This is the diplomacy of rich countries to establish their intellectual superiority. It is unfortunate that they have less interest to look into the problems of the third world.

The couriers of these pseudo-useful things are the multinational corporations and transnational corporations. In International arena the non concerning role of multinational corporations and transnational corporations which are dumping from one country to another is also a considerable issue while describing challenges related to international environmental issues. Multinational corporations have largely played a low profile role in international policy debates, preferring to await the formulation of national environmental policies and then responding. Environmental legislation plays a minor role in influencing MNC decisions about foreign investment and the choice of location for overseas operations.

In several ways Environmental damages are also caused by political conflicts and militarization prevailing time to time in several regions and countries of the world. The military conflict due to political tensions in the Central American Countries, the Middle East and several Asian nations including India, Pakistan, Afghanistan and Sri Lanka is also causing severe damage to the human environment. The activities of Terrorism specifically after 9/11 attack on USA and prospective dangers of bio terrorism are posing dangers of severe environmental damage. It is well known that wars and conflicts not only damage the natural environment but also the social fabric of the affected population which is hard to reconcile. The Gulf War over Kuwait in 1991 destructed large oil spills and extensive fires in 613 oil mills of Kuwait which burnt about 4 to 8 million barrels of oil everyday and caused massive clouds of smoke and toxic gaseous emissions. Nearly 1-2 million tones of green-house gas carbon dioxide emitted everyday, together with amounts of sulfur and nitrogen oxides, carbon monoxide and organic compounds. It not only damaged natural resources but also world peace and harmony.

Militarization has diverted considerable human resources meant for human development to unproductive and rather destructive means and purposes. One more problem needs attention that how the process of militarization is causing harm to developing nations that they in their national budget, instead of allocating money to rural development schemes, are contributing a large part of the budget in procuring arms and ammunitions. Since 1970 the world has spent approximately US $ 17 million on military activity. Politicians in India and several other developing nations on the contrary believe in spending 45 percent of annual national budget on making and buying better arms. The competitive arms race is on account of the bare subsistence of the rural masses in the developing countries, which is leaving huge amounts of debts for the future generation to pay. This military advancement is also a premonition of the lurking dangers on flora, fauna and personae.

The socio-political insinuations of biodiversity use and conservation in developed vis-à-vis developing countries is another relevant perspective in international environmental politics, which is decisively influencing the biodiversity of third world countries. Biodiversity of developing nations is greatest treasure not only for them but for the whole world. The agricultural scientists and environmental scientists visualize that the future of world economy and trade will depend on the biodiversity of the south Asian countries because of the capability of the soil to produce wide varieties of crops, cereals during different seasons. They opine that in future these biologically rich countries will cater to the needs and demands of the world. They can use it as 'political weapon' for political bargains with the developed world. But the biggest hurdle in front of the 'gene-rich' developing nations is that they need political solidarity in order to put demands. Several gene-rich countries including India share a considerable part of its biodiversity with neighboring countries in south and south-east Asia. Our natural ecosystem extends over and meets the political boundaries of other adjoining countries – such as the Sunder bans mangroves across Bhutan and other Himalayan ecosystem across Nepal. Several species of plants are found in south-east Asian countries and those occurring in the eastern Himalayas are found in Myanmar and China. In this situation, the question arises that which country singly would receive the benefit.

Biodiversity gives 'sovereign rights' to the poor developing countries of the South on their tropical forest and its genetic resources within their political frontiers. This means that the south nations including India and other countries of the sub-continent can legitimately claim fair returns for the use of their biodiversity by the developed nations. In the last few years the colonial and neo-colonial powers in the rich countries of north particularly have been rewarding them in the field of food and medicine. But in order to protect the weaker nations of the South from becoming an experimental ground for the North's genetically engineered monsters there is the need of an international legal and political instrument in the form of convention called 'Bio-safety Protocol'.

There is a stressing demand to develop a concept of environmental security by the political scientists and scholars of international relations across the world, who started thinking seriously about the challenges of the global environment. They argued for a new paradigm focusing on the idea of 'comprehensive security' which includes national security along with environmental security. They are of the opinion that the existing concept of national security is incomprehensive unless it is supportive of environmental issues. Thus the old paradigm of security lays emphasis upon military build up and thereby defending the country's internal and external sovereignty; while the new paradigm of security encompasses the impact of environmental degradation since it affects the security of all states, irrespective of their eco-strategic location and level of advancement.

Against the said background, several political agreements and instruments in the form of treaties, conventions and protocols were signed between countries at the international level for the protection of endangered flora and fauna and banning of trade in wildlife, for the protection of stratospheric ozone cover, for reduction of global carbon emission to avert climatic change; for conservation of global biodiversity; to combat desertification; to pervert pollution of seas and several other environment problems.

International environmental concern dates to the 1960s. Some of the important international political agreements to save the global human environment are as follows:

1. The Antarctic Treaty (Washigton,1959);
2. The Convention on International Trade in Endangered Species of Wild Flora and Fauna (CITES) (1975);
3. Vienna Convention for the Protection of Stratospheric Ozone' (Adopted March 1985);
4. The Montreal Protocol on the Substances that Deplete the Ozone Layer(1987);
5. The Basel Convention on the Control of Transboundary Movements of Hazardous Wastes and their Disposal 1989;
6. The United Nation Convention on Climate Change 1994;
7. The Third Conference of the Nations to the UN Convention on Climate Change, Kyoto, Japan, December 1997;
8. Convention on Biological Diversity 1992;
9. United Nation Convention to Combat Desertification 1993;
10. South East Asian National Agreement(ASEAN) to safeguard essential Ecological Processes and Life Supporting Systems;
11. Convention on Long Range Trans Boundary Air Pollution;

12. The Hague Declaration 1989 by representatives of 24 Nations including G7 Nations to combat any further global warming of the atmosphere, Convention on the Wetlands of International Importance especially as Water Fowl Habitat (Ramsar,Iran 1971) Convention on the Conservation of Migratory Species of Wild Animals, (Bonn, Germany, 1979);
13. London Guidelines on Chemicals in International Trade (1994);
14. Convention Concerning the Protection of the World Cultural and Natural Heritage (Paris, France, 1972);
15. Berne Convention on the Conservation of European Wildlife and Natural Habitats (Switzerland, Sept. 1979);
16. Paris Convention for the Protection of Birds useful to Agriculture (Paris, March 1902);
17. The Convention on the Conservation of Antarctic Marine Living Resources (Hobart, Tasmania, 1980);
18. The FAO International Undertaking on Plant Genetic Resources (Rome, Italy, 1983);
19. Lusaka Agreement on the Prohibition of Illegal Trade in Wildlife, (Zambia, 1994);
20. International Convention for the Regulation of Whaling (Washington, 1946);
21. International Convention on the Protection of Birds (1950);
22. Convention on the Protection of Marine Environmental of Baltic Sea (Helsinki, 1974);
23. Barcelona Convention for the Protection of the Mediterranean Sea Against Pollution (1976);
24. United Nations Convention on the Laws of Sea (1982);
25. Convention on the Prevention of Marine Pollution from Land Based Sources (Paris, 1974);
26. The International Convention for the Prevention of Pollution from Ships (London, 1973);

27. Convention on the Prevention of Marine Pollution by Dumping of Wastes and other Matters (London, 1972);
28. The International Convention on Oil Pollution Preparedness Response and Co-operation (1990);
29. Convention for the Protection of Alps (Salzburg, 1991);
30. Stockholm Convention on Persistent Organic Pollutants (POPs) (May, 2002);
31. Rotterdam Convention (For certain Hazardous Chemicals in International Trade) (Ratified on 24th May, 2005);
32. Cartagena Protocol on Biosafety India has ratified it on January 17, 2003.

There are some other International Treaties/Conventions/ Declarations which are as follows:- Agenda 21, Convention to Combat Desertification, Prior Informed Consent (PIC), Intergovernmental Panel on Climate Change (IPCC), Convention on International Trade in Endangered Species of Wild Fauna and Flora (CITES), World Trade Agreement, Helsinki Protocol to LRTAP on the Reduction of Sulphur Emissions or their Transboundary luxes by at least 30 percent, Sofia Protocol to LRTAP concerning the Control of Emissions of Nitrogen Oxides or their Transboundary Fluxes (NOx Protocol), Geneva Protocol to LRTAP concerning the Control of Emissions of Volatile Organic Compounds or their Transboundary Fluxes (VOCs Protocol), United Nations Conference on the Human Environment, United Nations Conference on Environment and Development (UNCED), Framework Convention on Climate Change (FCCC), Convention Secretariats of the UNEP.

The principal players in the management of international environmental issues have been national governments, the scientists working with the United Nations Environment Programme (UNEP), and national environmental groups. Whereas policy decisions are made by state actors, scientists have been actively involved in explaining the nature of environmental problems to decision makers because of their lack of expertise in these issues.

The international political system had a multilateral responsibility for the environment and as a realization the United

Nations Conference on the Human Environment (UNCHE), held in Stockholm, Sweden, in June 1972, was the first world conference that brought together delegates from 113 countries from the North and South to discuss environmental issues. This Conference was held in the era of Cold-war between communists (USSR) and capitalists (USA) blocks. The USSR and Eastern European countries boycotted the conference. The conference adopted a Declaration on the Human Environment, with 26 principles and 109 recommendations for action.

UNEP i.e. United Nations Environmental Programme was established the following year in Nairobi, Kenya, in order to "serve as a focal point for environmental action and coordination within the UN system." Since then UNEP is engaged in a wide variety of activities. It has sought to encourage environmental consciousness with its objective of dealing with the fundamental causes of environmental destruction, rather than merely arranging for clean up. Its major accomplishments have been in the areas of developing international law and promoting environmental monitoring. Under UNEP's auspices, regional pollution control agreements were adopted for eight regions of the world, Under the Regional Seas Programme, more than 120 countries and agencies participate, drift net fishing was banned and treaties have been drafted on biodiversity and global warming.

UNEP has also promoted development of a body of "soft law", a set of international guidelines and principles for environmental management which may contribute to the "gradual build-up of consensus among governments, in the field of international environmental protection". Principles of conduct were drawn up for shared natural resources, weather modification, offshore mining and drilling, a world charter for nature. It banned and severely restricted chemicals, marine pollution from land-based sources, and tried to introduce environmentally sound management of hazardous wastes, environmental impact assessment, and exchange of information on chemicals in international trade. A Global Plan of Action for Marine Mammals to provide guidelines for the management of coastal zones to protect marine mammals was also made. It has been providing training in environmental management and monitoring and publishing technical guidelines and manuals in cooperation with UN bodies and other organizations. These supply scientists, technicians, and planners with the necessary information to carry out environmental assessment and management.

UNEP has assisted other United Nations bodies with the drafting of action plans and coordination of international action in the areas of tropical forests, wildlife preservation, and biological diversity. FAO, UNEP and UNESCO completed a Tropical Forestry Action Plan. In 1980 the major multilateral development organizations, including the World Bank, adopted the Declaration of Principles on the Incorporation of Environmental Considerations in Development Policies, Programmes and Projects. UNEP has also had some notable failures too. The 1977 UN Conference on Desertification, for example, charged UNEP with coordinating its Plan of Action. Its efforts to elaborate the concept of "eco-development" were also cut back in order to focus on practical ways to infuse development projects with a concern for environmental side-effects.

Despite demonstrations of good will toward the environment, there remain profound political disagreements between countries about the priority that should be accorded to environmental issues and the particular measures that should be collectively pursued. International cooperation is made difficult by the general nature of the international political system, as well as by specific differences between countries' interests, particularly between developing countries and developed countries.

Concrete differences exist between countries about how to protect the environment and how much environmental protection is desirable. Some of these North-South differences were defused, or at least temporarily set aside, at a small high-level meeting of development experts from the North and South held in 1971 at Founex, a small town outside Geneva, in preparation for UNCHE. The North has not been monolithic, however. The United States took an early leadership role in international environmental issues through domestic and foreign measures. Since 1987, much international discussion has been shaped by the report of the Commission on Environmental and Development (the Brundtland

Commission), *Our Common Future,* which addressed the questions of sustainable development and the need to maintain economic development in the Third World while ensuring environmental protection. UNEP tried to generate consensus through use of a similar strategy in each case, but the cases differed significantly in their reception of the UNEP initiatives.

Yet there are genuine expectations from UNEP to check and control the deteriorating environmental and ecological scenario of the world. It is also required that the arbitrary role of the nations like USA on environmental issues whether taken up by UNEP or outside other organizations be it WTO ,G-8 etc, should be in consonance with the global environmental security. With the implementation of the policies of globalization , liberalization and market economy, commonly agreed new international environmental initiatives need to be taken up to sustain biotic environment for the survival of all species of the planet.

Reducing global warming by cutting carbon emission preventing ozone depletion by searching substitute for CFC, toxic waste reduction and its safe disposal, conservation of biodiversity and protection of species from extinction, arresting soil erosion and desertification will command major political decisions and diplomatic attention and have to be fought by the international community. European and Latin American countries can work together to save the region's deteriorating forests; nations bordering the Baltic sea could join together to reverse its degradation; countries in the Indian sub-continent i.e. India, Nepal, Myanmar, Bhutan and Bangladesh could combine forces to reforest the Himalayas, avert soil erosion, conserve biodiversity and reduce the frequency of crop-damaging floods, and new North South alliances can save migratory birds. And these are several environmental issues of global significance which require political solution.

The role of Nongovernmental organizations (NGOs) involved with environmental and developmental issues have also sought to publicize international environmental issues and to pressurize their respective governments to undertake more active efforts for environmental protection. Taken together, these non-state actors have played important roles in environmental agenda-setting and in improving understanding about the extent and nature of environmental threats, as well as in actively formulating and promoting specific policies and arrangements for environmental protection thereby affecting domestic programmes as well as foreign policies.

The National Perspective begins with the environmental consciousness that gets reflection in the ancient Indian scriptural note:

> Fire is his head, his eyes are the moon and the sun;
>
> the regions of space are his ears, his voice the revealed Veda;
>
> the wind is his breath, his heart is the entire Universe;
>
> the earth is his footstool, truly he is the inner soul of all.
>
> —*Mundaka Upnishad*

The Divine is described through natural phenomena in Indian philosophical and religious tradition. Since Vedic ages India has been a nature worshipping country and nature is regarded 'The Mother' the ultimate source of creation. The belief in care for nature and the natural creations are given prime importance in India.

Despite the existence of the noble thoughts about environment, with the rapid growth of industrialization, population explosion, unplanned development and excessive exploitation of natural resources etc environment deteriorated which initiated ecological debates and discussions at various socio-political platforms. The new economic policy of India is considered to be aggravating the problems of environmental protection and rights of the local communities to environment. While dealing with the political dynamics of environmental issues in the Indian perspective, the present article not only highlights the constitutional and legislative arrangements, judicial and human rights perspectives but it attempts to study the operational aspects of eco-dynamics of the system also. Before discussing constitutional and legislative measures in this article it would not be improper to mention here that some of them in different perspective have already been discussed in this book in detail so it will be relevant here to profile in brief about this aspect which directly or indirectly has an impact on environment.

In nineteenth century during British rule in India, the legislation passed to protect wild life and water pollution etc had a narrow purpose with a limited territorial reach .The Indian Penal Code of 1860 imposed a fine on a person who voluntarily "fouls water of any public spring or reservoir". Similarly The Indian Easements Act of 1882 was made to protect riparian owners against 'unreasonable' pollution by upstream users. Further Indian Fisheries Act of 1897 penalized for using explosives to poison the water causing the killing of fish and Indian Port act of 1908 to regulate the discharge of oil into port water and India Forest act of 1927 prohibiting the pollution of water in the forests etc were important legislation in pre-independent India. To control air pollution legislations came into existence were Bengal Smoke Nuisance Act of 1905 and Bombay Nuisance Act of 1912. In the area of wild life protection the Hailey National Park Act was enacted in 1936 under which Hailey Chow Jim Corbett National park was established in Uttar Pradesh.

The Indian Constitution recognizes the indispensability of and dependability on natural phenomena. The Constitution of India is distinct in the sense that it is amongst the fewer Constitutions of the world, which proclaim for the protection of the environment. The commitment of the Constitution towards the promotion and preservation of rich natural resources gets a clear and strong reflection in the Provisions of Directive Principles of State Policy and in the Fundamental Rights and Fundamental Duties, which in turn have been strengthened by judicial interpretation. Thus it is evident that the courts have also granted recognition to the right to a wholesome environment as being implicit in the fundamental right to life. It has also stated that it should be the fundamental duty of every citizen of India to protect and improve the natural environment including forests, lakes, rivers and wildlife, and to have compassion for living creatures.

The Constitution of India lays emphasis on the protection of environment through Article 48-A of the Directive Principles of State Policy, which ensures that "The state shall endeavour to protect and improve the environment and to safeguard the forests and wildlife of the country." The 42nd amendment also expanded the list of concurrent powers as it introduced a new entry, 'Population control and Family Planning while 'Forests' and 'Protection of wild Animals and Birds' were moved from the state list to the concurrent list and as a result of this, parliament concentrated to legislate on environment and Article 51-A(g) was added through 42nd Amendment to the Constitution in 1976 which enshrines a fundamental duty of the citizens to environmental protection. The 73rd Constitutional Amendment Act of 1992 on revitalization of panchayati raj in Indian political system adds XI Schedule to the Constitution. It has eight entries which are linked to environmental protection and conservation. The panchayats, accordingly, are assigned important environmental functions, including soil conservation, water management, watershed developments, social and farm forestry, drinking water, fuel and fodder, non-conventional (like biogas) energy sources and maintenance of community assets which are significant from the point of view of environmental management. The revival of Panchayat Raj in India also envisaged increasing local initiative and participation including conservation of local environmental resources.

Article 253 of the Indian Constitution empowers parliament to make laws implementing India's international obligations as well as any decision made at an international conference association or other body. As the broad language of Article 253 suggests that parliament has the power to legislate on all matters linked to the preservation of natural resources. Parliament has used its power under article 253 to enact the Air (Prevention and Control of Pollution) Act of 1981 and the Environment (Protection) Act of 1986.

There was a growing awareness about environmental degradation after the Stockholm Conference on Human Environment 1972 and as a result a series of legislation passed and came into force. The first important environmental legislation act in India was Wild Life Protection Act 1972. The constitutional provisions have been translated into the "regulatory" laws that aim at environmental protection, exemplified under the umbrella of Environment (Protection) Act of 1986. The whole legal framework provides more specific laws. Some important amongst them may be mentioned as Water (Protection and Control of Pollution) Act 1974 ,Air (Protection and Control of Pollution) Act 1981, Forest (Conservation Act) 1980, Environmental (Protection) Act 1986,

Wildlife (Prevention) Act, 1972, Manufacture, Storage and Import of Hazardous Micro-organisms and Genetically Engineered Organisms or Cells Rules 1989, to regulate storage, use, trade, transport and disposal of hazardous wastes etc.

The Government of India started giving shape to a detailed 'promotional' policy framework comprising the National Forest Policy, 1988, the National Conservation Policy Strategy and Policy Statement of Environment and Development, 1992 and the Policy Statement of Abatement of Pollution, 1992. These policy statements complement the National Water Policy, 1992, the Forestry Act, 1948 and other related legislations having a bearing on natural resources and economic activities. Some sector policies such as the National Agriculture Policy 2000, National Population Policy 2000and National Water Policy 2002 have also contributed towards environmental management. National Environmental policy 2006 is intended to be a guide to action in regulatory reform programmes and projects for environmental conservation and review and enactment of legislation by the central, state and governments working at local level. This promotional framework of policies is operationally, financially and institutionally supported by numerous schemes of financial assistance, funds earmarked for specific schemes, and fiscal incentives in areas of social forestry, forestry, and wildlife protection and pollution control.

The Ministry of Environment & Forest (MEF) is primarily concerned with the implementation of policies and programmes associated with the conservation of the natural resources of India including its forests and wild life, lakes and rivers, its biodiversity, assuring the welfare of its animals and prevention and abatement of pollution. These objectives are aimed to be fulfilled through environmental impact assessment, eco-regeneration, assistance to organisations implementing environmental and forestry research education and training, dissemination of environmental information, creation of environmental awareness and international cooperation. Working for sustainable development, the ministry also serves as the nodal agency in India for the United Nation's Environment Programme(UNEP),South Asia Cooperative Environment programme (SACEP) , International Centre For Integrated Mountain Development (ICIMOD) and for the follow up of the United Nations' Conference on Environment And Development(UNCED). On matters pertaining to environment, the Environment Ministry is also entrusted with the issues relating to multilateral bodies namely the Commission on Sustainable Development (CSD), Global Environment Facility (GEF) and of regional bodies like Economic and Social Council For India, Asia & Pacific (ESCAP) and South Asian Association for Regional Cooperation (SAARC).To protect and preserve the environment, there are sets of regulatory and legislative measures. Besides these legislative measures, a National Conservation Strategy and Policy, 1988, a Policy Statement on Abatement of Pollution 1992 and National Environment Policy 2006 have also been evolved.

An Operational analysis of the eco-dynamics of the political system in India reveals the fact that despite the legislation of various environmental policies and programmes their implementation and execution remains unfulfilled due to the vested political and economic interests, bureaucratic hazards on the one hand and poverty, unemployment ignorance and lack of public awareness on the other. On environmental matters it is seen that there are differences of opinions between formal government institutions and structures such as interest and pressure groups, industry and environmental NGO's and local communities. There are also different tones of centre and the states under federal structure. With the coming of globalization and liberalization in nineties the challenge emerging before the government is how to manage the larger economic environmental constraints set by external forces like multinational corporations or foreign corporate power and employment generation offered by new industries (even though polluting) and development projects like the big dams or huge construction programmes.

Environmental policy and management in India took a new turn when Prime Minister Indira Gandhi, the then Prime minister of India, participated in United Nations Conference on Human Environment in 1972 at Stockholm. The protection of environment became a major element in the acts of planning and development. The due concern of enhanced quality of life and preservation of environment were focused upon together in the fourth plan (1969-74). The fifth plan (1974-79) and the sixth plan (1980-85) specifically included the concern of environmental protection along with planning and development and launched several programmes which incorporated the environmental norms and conditions in development projects, for example, during 1979-80 the appraisal of thirty hydroelectric and irrigation projects and fifteen thermal power generation schemes from the environmental protection view point.

Again in the early eighties during the regime of Mrs. Indira Gandhi the then Prime Minister, some legislative, administrative steps were initiated to ensure environmental protection. Mrs. Indira Gandhi executed her eco-ideas by supporting projects like WWF-I or Projects Tiger. Earlier in 1976 also, the subject of 'forest' was added to the concurrent list of the Constitution and Forest Consummation Act 1980 was passed which prohibited the use of forest land for 'non-forest purposes'. The central government constituted a committee chaired by Narayan Dutt Tiwari to recommend appropriate legislative measures and administrative machinery. It was decided that the formation of Department of Environment and Forests was necessary to protect environment and to make provisions for sustainable development. This department later on was turned into Ministry of Environment and Forests (MEF) under Rajeev Gandhi's Government. Since then Ministry of Environment and Forests has been working in cases of friction between business-industrial interests on one side and environmentalists/human rights activists on the other. The Ministry of Environment & Forest (MEF) designed a new legislation on forest titled "Conservation of Forest and other Ecosystem Act". But in practice the proposed Act was not officially available to public till mid 1995, but within the stipulated time it was leaked and the proposed forest conservation Act was met with organized resistance and informed criticism of the forest dwellers, tribal and the activists throughout the country.

During the eighties, National Drinking Water Mission with the initiatives of Rajiv Gandhi, the then Prime Minster of India started to provide drinking water to villagers by digging more wells through water resource management schemes. Some other important actions were also taken such as Ganga Action Plan in 1986 for cleaning up Ganga water, yet the plan got slowed down under bureaucratic procedures. National Wasteland Development board

(NWDB) in 1985 under Kamla Choudhary to green India's vast tracts of wastelands and to utilize five million hectares of land for fodder and fuel food plantations despite young prime minister's efforts did not succeed completely. Rajiv Gandhi also introduced panchayti raj institutions to further the socio-environmental programmes. It was a challenge as well as an opportunity to the decentralized democratic system to guard whatever environmental resources are still available and even regenerate the degraded resources like the land, village forest and water resources in and surrounding villages. But unfortunately in practice the initiatives taken by the government could not succeed in the desired direction due to various socio-economic and political reasons.

An important aspect of the operational dynamics of environmental issues in India is related to environmental clearance, which prior to the establishment of new industry in the state is Mandatory. Even NOC-(No Objection Certificate) for certain trade and industries from the state as well as MEF is necessary. A 1994 Modification made EIA-Environmental Impact Assessment statutory for many sectors such as industries, irrigation, power, transport, tourism which are in turn associated with ministries of commerce, energy, finance and others therefore if there is any tussle on ministerial level it is the Prime Minister who takes final decision on all 'contentious' issues like Enron, Cogentrix, or river projects like Cauvery, Upper Krishna and Tehri. The status of the heading person of MEF is only that of Minister of state resulting in curtailing the power of MEF. Thus any Cabinet Minister with Prime Minister can uproot the decisions of MEF. Earlier also there have been tussles between other departments such as power industry, agriculture, commerce and MEF. In V. P. Singh's government Maneka Gandhi, Minister of state for Environment and forest and Nilamani Routray, Cabinet Minister for Environment and Forest had tussle on eco-hostile industries proposal. She tried for the establishment of environmental court for 'Monitoring and speedy disposal of cases against polluting industries and contentious cases but all by herself. There are many such cases where this Crusading Minister found herself isolated for example–Sardar Sarovar Project (Baba Amte and Medha Patekar, a noted environmentalist), Narmada Dam Project during Prime Minister Deve Gowda's government, Many Lease proposals etc. Such an example is Silent Valley controversy in the early 1980's when DigVijay Singh was the Deputy Minister in the Department of Environment and Indira Gandhi Prime Minister. The rift was solved by Prime Minister in favour of Anti-Project environmentalist. U.P. government's Vishnuprayag Dam Project, adjoining the Valley of flowers in Garhwal region was stopped by the Prime Minister Indira Gandhi.

During the nineties, with the beginning of globalization, liberalization and privatization, India also introduced a new economic policy under which a new phase of rapid development began, unfortunately not in consonance with sustainable development. The then environment and forest minister, Kamal Nath putting forth the two million hectares of deforested land on lease to the corporate sector with the claimed proposition that the act would generate funds for the expansion of forest cover and the problem of the encroachment on forest level would be solved. But this plan, under the continued pressure of tribal and supportive NGO's was not revived by the latter environment and forest ministers. Jai Narain Prasad Nishad in the United Front Government in September 1996 made it clear in his statement, "The local people who depend on forest for their livelihood should be the ones to benefit from this activity, not industry."Under the stewardship of Narsimha Rao's government drafted the forest bill in 1992 and Environment Impact Assessment (EIA) notification in 1994, made appraisal of new projects of over 50 crore in 29 sectors by experts necessary. The bill for establishing environment tribunals introduced by Narsimha Rao's government in 1995 but received widespread criticism for it exempted the governmental agencies from its purview and it did not provide localites the right to information of natural hazards.

Various studies on environmental scenario in India make it clear that in denotifying various sanctuary areas for industrial development, environmental considerations were relegated into background by the politicians, which also generated fight between environmentalist and the government many times. In Himachal Pradesh the 'Darloghat Santuary' was denotified for the construction of cement plant; in Gujrat the 'Narayan Sarovar Sanctuary' was denotified in 1993 for Lignite mining and reduced the area of sanctuary from 765.79 sq. km. to 94.97 sq. km. and in

Maharashtra the 'Melghat Wildlife Sanctuary' was denotified for industrial development. The 'Konkan Railway Project' in the Western Ghats has been cleared by the government of India. It would destroy the flora and fauna of the surviving rainforests of Western Ghats

An analysis of the party system in India reveals the fact that environmental issues have been overlooked more or less by all the political parties in India, be it national or regional. Environmental concerns began to take place in the election manifestoes of different political parties in the mid eighties. The election manifesto of Mr. Rajiv Gandhi, the then Prime Minister of India, which declared that Congress party "will optimize use of natural resources so that while meeting the current needs of growth, the resource base is oriented towards sustainability." Similarly in 1989 Janta Dal Election manifesto stated "with the majority of the country people still being poor, the degradation of environment is leading to the degradation of resource base with which alone the basic needs of people can be met." The Bhartiya Janta Party, in its manifestoes considered protection of environment as a 'sacred responsibility' as it declared, "our land, air and water, our hills , forests, flaura and fauna – all are sacred." But ironically the suggestions for environmental problems adopted by the parties were never genuinely implemented by themselves during their regime.

The subject of environment sometimes becomes a matter of dispute and constraint between centre and states because it is always the centre that keeps an upper hand in sanctioning and giving approval to state projects by Ministry of Environment and Forests. The centre, through its ministries and governments corporations, often establishes large industrial projects in the states. Silting of these projects is usually a political matter with several states vying to locate the project within their own territory. Once a location is decided, the expected economic boost and prestige that these projects bring to a state and its chief minister, coupled with an implicit central government's threat to relocate the project if the state raises any awkward questions, influence most state governments to overrule the objections raised by their own environment and planning agencies.

With the objectives of monitoring, reassessing and commencing all sorts of environmental plans and projects operated in states, disseminating environmental education and creating awareness, making critical appraisal of all proposals, incorporating safeguards in industry and corporate development projects, and to set bilateral relation with the Department of Science and Technology and NCEPL at the centre, with research organisation and educational institutions etc the environmental Boards were formed by the states. These boards were constituted on the recommendations of the NCEPC, on the pattern of the national committee, with the Chief Minister as its chairman. It was the initiatory step of states in the direction of Environmental Management. But the problem emerged before these boards is that the members of these boards being political persons lack scientific and technical expertise, so these boards remain ineffective. The state government could not provide expert staff and infrastructure also. Thus the target of safeguarding environment remained unfulfilled barring a few states like Karnatka, Tamilnadu, Kerala, Gujrat, Maharastra, Andhra and Uttar Pradesh, which commenced few activities.

The operational analysis of the political system of India makes it clear that the working pattern of the political and administrative agencies is that of eco-culprits. More or less, all political parties are factually insensitive to eco-issues. The environmental policies that the state has adopted and the laws made to protect our land, forests, water, air and such resources and the power structures made to enforce or monitor them, have been violated and flouted by the entrenched commercial, industrial and forest interests working under the umbrella of political and bureaucratic system.

In this context, it is the judiciary which has adopted an increasingly active role to respond to the nascently upcoming cult of environmentalism and expose the political and economic gimmick of the government industrial lobbies and chastise the powerful culprits. In Indian democracy, the conception of environmental freedom has expanded from constitutional provisions to progressive phases of judicial decisions. Increasing examples of court decisions on environmental issues have put forth the significance of right to wholesome environment. Supreme Court's landmark decisions have sharpened the legal instrument of public interests litigation in the case of environmental protection. Thus it has strengthened the position of citizens on eco-litigation vis-à-vis the commercial and political polluters responsible directly or indirectly. The constructive role of judicial system is visible on environmental issues related to healthy life and survival of human beings. Along with environmental groups and public litigation activists, the judiciary has prevented the political system of India being unilaterally manipulated by political and partisan interests.

The judiciary turned into the saviour of eco-treasures in India in the early 1980s. The Supreme Court and High Court judges were referred to deal with public grievances over flagrant human right violations by the state or to vindicate the public policies embodied in statutes or constitutional provisions. This new type of judicial business is collectively called 'Public interest litigation' (PIL). The petitioner seeks to champion a public cause for the benefit of all society. This activist role of PIL judge is in contrast with its passive role, traditionally associated with judicial functions. Most of the cases of environmental actions in India are brought under Article 32 and 226 of the Indian constitution. The writ procedure is preferred to the conventional suit because of its speedy, relatively inexpensive and easy reach to the highest courts of the country. The Andhra Pradesh High Court in T. Damodar Rao and others Vs The Special Officer, Municipal Corporation, Hyderabad (AIR 1987 AP 171) observed that: "Protection of environment is not only the duty of the citizens but is also the obligation of the state and all other state organs including the courts. In that extent environmental law has succeeded in unshackling man's right to life and personal liberty from the clutches of common law theory of individual ownership." Thus Environmental litigation in India reveals the necessary proactiveness of courts in taking due and timely solutions for environmental safety and protection.

On the litigation for the conservation of forest the response of the Court shows a great concern for environmental protection which in turn reveals that the Courts have "lived up to the new moves and policy shift from revenue earning to sustainable use of forest resources." This shift is found evident in interpreting the concept of use of forest for non-forest purposes in connection with the requirement of prior approval of the federal government for the proposed action. Courts have already made it necessary that the holder of a mining license should not disturb without prior permission the natural growth of forests and this mandate is necessary not only in connection with 'reserved forests' but to all forests. In the case of Hyderabad Abrasive and Minerals Vs State of A. P. 1990, (Forest Law Times 119) it was noticed that the licensing does not provide any authority the power to vitiate the natural ecology and license already granted before the enforcement of the forest conservation law would be given renewal permission only with prior approval. Such cases show that the courts have understood that deforestation and resultant ecological imbalance have become a growing threat and further deterioration should be prevented. In April 1996 the Supreme Court has asked to stop any type of construction within 500 meters of the line of tide along the coastal line. This was a historical step for the protection of the country's long coastal area. Along with, the Supreme Court also gave a directive that India's coastal states should constitute 'Coastal Area Management Authority' to look after environment

Environmental pollutants can be seen as major hurdles challenging the social justice segment of the rule of law. In 1980 Justic V.R. Krishna Iyer held in Municipal Council, Ratlam vs Vardhichand, (Municipal Council Ratlam vs. Vardichand, AIR 1980 SC, 1622), that the demand for unpolluted environment is human right and this right should be provided despite the financial crisis of government. In the city of Ratlam, residents were suffering from bad smell from open drainage system, public excretion of slum dwellers and effluents from distilleries. They appealed through public litigation to the magistrate to take action to provide solace. The magistrate ordered the Municipal Council to remove the litters and nuisance generating pollutants. The council showed its inability in managing this on financial grounds. The Supreme Court tried to dig out in the law of public nuisance a social justice component of the rule of law and regarded its role very important in the case of environmental protection in the third world countries. The Court held that the operation and implementation of law can be against statutory bodies and others irrespective of any matter because human rights are observed by the state as superior to any budgetary monetary constraints.

Supreme Court has given a broad meaning to the Right to Life under Article 21 of the Constitution, which has been used in various in India. It includes the right to survive as a species, quality of life, the right to live with dignity and the right to livelihood. The Supreme Court has used the right to life as a basis for emphasizing the need to take drastic steps to combat air and water pollution. It has directed the closure or relocation of industries and ordered that evacuated land be used for the needs of the community. The courts have taken a serious view of unscientific and uncontrolled quarrying and mining, issued orders for the maintenance of ecology around coastal areas, shifting of hazardous and heavy industries and in restraining tanneries from discharging effluents. There are several cases illustrating all these.

The ambit of the constitutional Right to Live was expanded when Indian Judicial system which began to consider the right to clean environment as an inextricable part of this right. Through the decisions in the land mark cases it has accorded the right to life. In the first case of M. C. Mehta V. Union of India, (AIR 1987 SC 965 P 966) the Court observed that poisonous emissions from a factory posed significant questions on the scope of the right to life. In the third M.C. Mehta case (AIR 1988 SC 1097 PP 1047-1048) the concern for the life, health and ecology weighed against unemployment and loss of revenue when the Court gave decision for the closure of the polluting industry if they ignored the option of establishing treatment plants.

It is also interesting to note that the Supreme Court has never made the clear stance directly in any case that the right to safe and sound environment is embedded in the constitutional right to live. Nevertheless, the enthusiasm with which the courts issue decisions with the objective of protecting the lives and health of people, and the 'ecology', insinuates the obvious judicial recognition of such a right. This development is in consonance with the juristic view that even if the bill of rights of a constitution does not articulate concern over the quality of environment, a right can be deduced from the due process. The State High Courts have been more specific in openly verbalizing the right to clean environment .In Attakaya Thangal V. Union of India, the Kerala High Court observed, "The right to sweet water and the right to free air, are attributes of the right to life, for these are the basic elements which sustain life itself."

The fundamental Right to Life in Article 21 is supported by Court's conclusive direction to close the polluting tanneries in the Ganga pollution case, "we are conscious that closure of tanneries may bring unemployment, (and) loss of revenue, but life, health and ecology are of greater importance to the people". The High Courts have also explicitly recognized the environmental dimension of Article 21, while considering a writ petition to stop the Life Insurance Corporation and the Income Tax Department from raising residential houses in a recreational zone, the Andhra Pradesh High Court said, "There can be no reason why practice of violent extinguishment of life alone should be regarded violative of Article 21 of the Constitution. The slow poisoning by the polluted atmosphere caused by environmental pollution should also be regarded as amounting to violation of Article 21 of the Constitution." Judges of the High Courts of Rajasthan Kerala and Himachal Pradesh, too have observed that environmental degradation violates the fundamental right to life.

Right to livelihood is one of the direct facets of the right to life. The Supreme Court has upheld this point in the case of Olga V. Bombay Municipal Corporation. A government scheme evicting the pavement dwellers from Bombay was challenged by a writ filed by the Petitioners, a well known journalist and two pavement dwellers. The thrust of the petitioner's argument was that the right to life includes right to livelihood, and since the pavement dwellers would be deprived of their livelihood if they were ousted from their slum and pavement dwellings, their eviction would be similar to deprivation of their life, and was held unconstitutional. Thereafter, in Banawasi Seva Ashram V State of Uttar Pradesh, the Supreme Court protected the tribal forest dwellers who were being ousted from their forest land by the National Thermal Power Corporation Limited for the Rihand Super Thermal Power Project. In other such petitions, the Supreme Court directed interim orders to state agencies to resettle and rehabilitate tribal who were being displaced by dams like Tihari and Narmada projects.

Similarly the right to know stands for the poor and oppressed enabling them to inquire into the actions and activities of the government functionaries thereby checking the abuse of public office by the powerful hegemonic order of authorities. It has entrusted the public to challenge government policies or inaction that threatens the public health. In the case of Reliance Petrochemicals Vs Proprietors of Indian Express Newspaper, Justice Mukharji reading between the lines upheld that right to know stems from the right to life. In his words "We must remember that the people at large have a right to know in order to be able to take part in a participatory development in the industrial life and democracy. Right to know is a basic right (to) which citizens of a free country aspire (in the broadening horizon of the right to life) under Article 21 of our Constitution. That right has reached new dimensions and urgency."

By providing the right to know Courts have legitimized public's interest and capacity to retrieve the authentic information from government and public offices thereby establishing the notion of the government officials' accountability towards the people. It lays ruling that how far information can be withheld from public and with what measures a person can retrieve those confidential information. The proposition of a transparent government is an explicit emanation from the right to know which appears to be obliquely embedded in the right of free speech and expression guaranteed under Article 19 (1) (a) which states "This is the new democratic culture of an open society towards which every liberal democracy is moving and our country should be no exception."

But this right to know, also known as access to information, in environmental matters, is somewhat debatable. The access of public to environmental information and influence of public on decision making can bring in a "more equitable distribution of natural resources," which will curb/control to an extent the authoritative decisions of government. Example may be given here of the decisions of government regarding the selection of site for dams. Selecting a site can cause displacement of thousands of people and depriving them of their lifestyles and livelihood. A responsible government therefore is expected to bring out public notification for its river development plans and should receive public feedback.

Now we come to the debatable issue of environmental degradation and human rights which have a cognate relationship. Y.K.Sabharwal Ex. Chief Justice of India tracing this relationship has stated that how environmental degradation affects the human rights in several other ways. The overuse of natural resources may engender unemployment in villages and emigration to cities which affects the environmental ambience and social milieu in the cities and hence the exercise of basic human rights gets hampered. The distorted and polluted conditions generate infectious diseases in countries where even the basic health facilities are not available and therefore people are deprived of the right to a healthy life. As Justice Y. K. Sabharwal has remarked "From the 4,400 million of people who live in developing countries, almost 60% lack basic health care services, almost a third of these people have no access to safe water supply ... Environmental degradation worsens existing problems suffered by developing and developed countries. Air pollution, for example, accounts for 2.7 million to 3.0 million of deaths annually and of these, 90% are from developing countries." Next construction of dams and various other hazardous problems create environmental refugees who suffer from significant economic, socio-cultural, and political consequences.

Highlighting the close relation between environment and human rights Justice Sabharwal has also traced the commonalities between environmental and human rights law that enable the creation of a field of cooperation. Both disciplines have "deep social roots" but whereas human rights law is more rooted within the collective consciousness, environmental degradation is generating a new "environmental consciousness" and both are increasingly becoming internationalized. The international community has realised the importance of human rights as well as of the environment. From the Second World War onwards, the relationship between State and individual has become important to the international community and environmental degradation has also crossed political boundaries. Thus both are of critical importance to the preservation of world peace and security. The protection of the environment is internationalized and the State-Planet Earth relationship has become a matter of deep concern for the international community. Environmental and human rights law have universalized their object of protection. Human Rights are presented as universal and the protection of the environment appears as everyone's responsibility. Nearly all global and regional human rights bodies have accepted the link between environmental degradation and internationally-guaranteed human rights. In nearly every instance, the complaints brought have not been based upon a specific right to a safe and environmentally-sound environment, but rather upon rights to life, property, health, information, family and home life. Underlying the complaints, however, are instances of pollution, deforestation, water pollution, and other types of environmental harm.

In view of the above analysis of the eco-dynamics of the political system in India it may be observed that there are legislations and laws to protect the environment but lack of proper implementation of these policies is the essence of eco-centered challenges. It is suggested that the existing environmental laws in India need to be managed in the way so as to make them effectively and conveniently available to the people. Strategically well thought eco-development programmes and policies along with effective policy implementation, good governance, people's awareness and participation, and governments' efforts to check poverty, illiteracy and ignorance are some important steps in this direction. Media also can play a significant role in raising environmental issues, creating public awareness, checking and controlling the behavior pattern of governments and administrative machineries and finding ways for sustainable development.

There is a greater need to materialize the concept of eco-democracy to achieve the goal of sustainable development. For the making of a sustainable society, democratic system from grass root to top is essential as it provides opportunity for a variety of channels for negotiations and it also involves people's voices and active participation. Apart from creating a set of representative institutions, democracy also enables people's capacity to collectively resolve common problems like managing common resources. There is also a greater need for making policies for sustainable environmental management.

Elaborating the concept of sustainable development Justice Y.K. Sabharwal emphasised that there are three basic components of it. First is the precautionary principle, Second the principle of 'polluter pays' and third the principle of intergenerational equity. According to precautionary principle the state must anticipate, prevent and attack the cause of environmental deterioration. The principle of the polluter pays means that reparation has to be paid by the polluter for rehabilitation of the environment to its original state. The third assures accountability of each generation to conserve the diversity of the natural and cultural resource base.

Coming to the international arena it may be suggested that without giving due concern to environmental conservation and ecological security of the nations, the economic and political policies to boost economic growth and development and efforts to eradicate poverty in developing countries, would remain infertile. It is the high time that world's political agenda for directing and executing for development need to be ecologically oriented. Each and every government of the world should have a greater responsibility to fulfill the basic needs of the present generation and greater accountability towards future generation to meet their needs for their sustenance. No economic policy can be called successful if its prosperity comes at the expense of environment of coming generations and if its implications lead to the growth of poverty and deterioration in the environment.

It is true that till today the development strategies so far followed worldwide have neither done justice to the man nor to the environment. The ecological misconduct of men, it may be observed, brought economic prosperity for the few. Although the concept of ecologically sustainable development, judiciously merging economy with ecology, is evolving gradually but little has been achieved and much is to be done.

The political leaders of the developing countries could not succeed despite identifying to resolve the challenges generated by the nexus between population, poverty and pollution. An obsession may also be seen in the political leadership of developing countries for adopting western model of development irrespective of their different socio-cultural and economic background.

The states are the important actors in shaping environmental politics and deciding the future of the world as they sign agreements and treaties make efforts to implement them and also make arrangements to change the behaviour of those actors who violate rules. Thus, solutions to environmental problems of the world require a two way solution i.e. involving active coordination of the states whether developed or developing and coordinated action among all sections of civil society throughout the world.

Future of world much depends upon growing awareness and a genuine pledge to protect the environment. The policy makers and planners, bureaucrats and technocrats who make decisions about the development of a country need to realize and to uphold their eco-conviction with greater firmness and implement these in planning and governance. Environmental security thus should be the subject matter of serious attention of global politics and it requires a comprehensive, collective political approach to deal with problems and challenges. Political and economic experts should have a deep faith to promote ecological politics and ecological economics as well.

NOTES AND REFERENCES

1. Surender Singh Chauhan, *Environmental Protection & Management from Stockholm to Rio & After*, Kalinga Publications, Delhi, 2004.
2. Aftab Alam (Ed.) *Human Rights In India—Issues And Challenges*, Raj Publications, New Delhi, 2001.
3. Deepender Basu (ed.), *Environment and Ecology: The Global Challenge*, Printwell Publications, Jaipur, 1995.
4. Digvijay Singh, *The Eco-Vote People's Representatives and Global Environment*, Prentic Hall of India Private Limited, New-Delhi-1985.
5. Peter Borrelli (ed.), *Cross Roads Environmental Priorities for the Future*, Island Press, Washington D.C, Covelo CA, 1988.
6. Pravin Sheth, *Environmentalism Politics, Ecology and Development*, Rawat Publications, Jaipur and New-Delhi, 1997.
7. R. Guha, *Environmentalism: A Global History*, Oxford Publications, New-Delhi, 1999.
8. S. Krishna, *Environmental Politics*, Sage Publications, New-Delhi, 1996.
9. South Asia Politics, 'Review of National Environment Policy 2004, Institutions, Incentives & Communities in Environment' Vol. 4, No. 1, May, 2006.
10. Draft National Environment Policy, Ministry of Environment & Forest, Government of India, New Delhi, 2006.
11. Economic And Political Weekly, 'Climate Change: Need for New Economic Thought,' Vol. XLII No. 6, February 10 to 16, 2007.
12. Justice Y. K. Sabharwal, "Human Rights And The Environment",www. supremecourtofindia.nic.in
13. India 2007, New Delhi.
14. www.unep.org

10

Social Challenges to Environment and Administration

Dr. Renu Srivastava*

Introduction

Society is engulfing its own surrounding at a faster pace than the capacity of the nature to replenish itself. Impact accruing from the hi-tech industrialization and urbanization process has not only imperiled the ecosystem and environment beyond repair but it has jeopardized the existence of the human race itself. Needless to say that ecosystem is home and habitat for numerous species of plants and animal forms. Nevertheless, man's encroachment of rivers, mountains, valleys and forests has resulted in distorting the natural life style and habitation of these biological species.

How perilous environmental degradation and merciless denudation of forest cover could be for our own existence is exemplified by frequent accidents reported in the news like mad elephants masquerading the village hutments in the regions of Orissa, Kerala and Assam, Leopards let loose on the killing spree in Maharashtra and Gujarat. Similarly in the past five years in Varanasi and Mirzapur districts, people have become day-to-day victims of monkeys' assault. Why ? The answer lies in the mindless destruction of the well-thought out ecological order in the name of development

* Reader, Department of Political Science, Vasant Kanya Mahavidyalaya, Kamachha, Varanasi.

and urbanization. Therefore, society is responsible for desecrating the natural order causing excessive pollution of air, water, soil, and sound.

Men are limitlessly greedy unless the noose is tightened around their own necks. So the time beckons them now to have cuts on their extravagant life styles or perish. No wonder, high-sounding slogans of scientific revolution and technology seems to be taking a back seat in the 21st century.

The term environment has suddenly become very catchy with the scientists and governments. Let us analyse what is so significant about it. Literal meaning of the term denotes the physical atmosphere surrounding the mankind on planet Earth. Humans and environment are integral components of each other. All kinds of flora fauna, living beings and natural wealth are its part. Conservation of environment is essential for the future of posterity. However, normal course of environment is being affected because of human interference in the wake of modern materialistic and consumeristic culture, giving rise to various forms of pollutions as well as deadly disease and mental tensions. Therefore, if "welfare of all" is to be the aim, environment protection and effective pollution control drive must be given the top priority.

The cultural view: Contrary to the current perspective, the Indian view treats environment as a part of divine order and a reflection of God Amighty. According to Hindu faith human life consists of five natural forces—Air, water, fire, earth and sky which are considered the five major components of physical environments. They are governed by intrinsic law of interdependence and cannot exist in isolation from each other. (R. Singh 1993: 3).

(Singh 1999: 3) Similarly other natural forces as mountain, rivers, oceans and trees are not considered as inanimate resources but abodes of deities and Gods. River Ganga is believed to be granting salvation from sins. Neem. Mango, Tulsi, Banyan and peepul are not mere trees but considered holy and sacred. Animals and birds are carriers of Gods and Godesses, and the earth as the mother of all. Through ages and millennium the culture of nature worship continues to inspire the knowledgeable and enlightened who strongly believe that physical form is nothing, if it is devoid of spiritual content. Hinduism prescribes spiritualism as the panaceas of all evils. Surprisingly, several western scientists too corroborate the mystic nature of the Universe. They look upon the world as a KOSMOS which is named as implicate order. "The essential features of this implicate order are, as we have seen that the whole universe is in some way enfolded in everything, and that each thing is enfolded in the whole" (D. B. Ohm and B. J. Hiley 1993: 382). There is an order enfolded in the very process of the universe but that unfolded (or implicate) order may not be readily apparent (Seyyed Hossein Nasr 1996: 149).

Technological View

Scientific analytical method destroys the basic unity underlying the diverse forms by treating nature into separate compartments. It treats nature as a fountain of constantly flowing resources. However, science and technology have influenced every aspect of human life. It is indeed an unprecedent blessing to human kind, ever since it was invented four hundred years ago. Scientific age is characterised by a boom of scientific devices which have the potential to squeeze out even the last drop of energy existent in natural resources. At a glance, to enumerate a few, quarrying of rocks and mountains to procure building materials and concrete, mowing down trees and forests for wood work and furniture, damming the lakes and river waters to generate electricity, boring underground water for drinking and irrigation, mining coal and churning sea water for extracting fuel and petroleum, extensive mining of ores and minerals to run industries and missile programmes etc. are definitely noted achievements of scientific inventions. Hence science, must be credited for enhancing prosperity and over simplifying the day to day life of human kind, but for its most hazardous consequence-enslavement of nature by humans.

The rat race of large scale production is sapping out the fertility of earth and inviting new diseases and dangers. Rivalry in world trade and commerce, liberalization and free trade all lead to fear and imbalances which converts into war. As very rightly pointed out by Anand Coomaraswamy that "world trade and world wars are cogeneric evils." (Quoted in Reghubir Singh 1999: 9).

We have developed nuclear weapons and missiles in the name of development and higher materials standards of life. knowing fully well that this development is strictly confined to a small elite section of society. Technological innovations support the economic empire of the rich by simultaneously marginalizing the poor. Entire concept of development is based on the exploitation of countryside resources, agriculture, forest and coastal areas which are exclusively inhabited by the indigenous and rural people who have no access to modern technology. Illiteracy and poverty renders them incapable of learning any new mode of communication while contemporary development is strictly based on modern information technology. For example computers, internet and digital forms of communication are restricted only to very well of and educated who are few in number. In this way environmental degradation and pollution are caused by the action of rich but the consequences are suffered by the grassroots and the poor. In a nutshell we can say that development in not accessible to all. Moreover, the present concept of development is technology based which is anti-environment. Is it possible to invent environment friendly technology? Perhaps yes because technology is essential for development. And as we have become dependent on tech-induced development we cannot go back. Subsequently, we are playing with the idea of sustainable development. Just as "development" was the key word of pre-liberalization era "sustainable development" is the key term of post globalization era.

Of late, sustainable development has replaced the idea of unbalanced and unlimited development. It means development as well as conservation for the posterity. Unfortunately, sustainable viewpoint does support restricted development but does not redefine it. It does not set any new framework which can be holistic and inclusive of all. It remains confined to economic and industrial growth of profit and loss. A further improvement in the concept in taking place in form of human resource development on the model of China and Japan which again is no better than technical development because it is devoid of any ethical or moral content. Eventually human development ends up into physical and economic development only (op. cit., Singh) on the contrary, Vaidik culture treats nature as sacred and venerable, not only as mere resources. Much water has gone down the Ganga ever since the race of development started and now it has been converted into a dirty sewage as a result of development administration.

Environment Crisis

In a wider connotation environment crisis is not only a crisis related to atmospheric pollution but it is basically a crisis of modern civilization and culture. Scientism and mechanical nationalism have become the tools of conquering nature and interfering with the laws of nature as a result of which natural balance of universe or implicate order is endangered. (R. Singh 1999: 7). In order to fulfill his diverse needs man experiments with nature and makes undue interference into its method giving rise to artificiality and pollution. For example nuclear detonation at one place affects the whole space and time.

Ozone depletion, acid rain, global warming, floods and frequent earthquakes are the ramifications of disbalancing the micro relationship between the nature and its various organs. Approach of industrialization and mechanization for development has threatened the pristine relationship of man with nature. Startling researches in the field of nuclear and genetic engineering seem to suggest that science has succeeded in recreating a new human and physical system. However, it is oblivious of the dire consequences it may have to face in future by producing and eliminating desired humans and animals. Can the man become the creator himself ? It is a troubling question. If we look at the changes induced by science into agriculture and industry, we find that we have only damaged the environment instead of benefiting the weak and the poor the larger section of humanity. Ever increasing use of fertilizers, pesticides and insecticides are turning the soil infertile. Similarly, over consumption of modern gadgets like plastic, foam, air-conditioners, refrigerators, automobiles are polluting air, water land and forest beyond remedy.

Hi-Tech Development

The rat race of development amongst nations has blinded the government to its damaging consequences on environment. Ever since the industrial revolution, major advances in the standards of living of the European and Asian people were attained. After second world war, science and technology were sparingly used by developed countries to enhance production and national income. Suddenly technology took over economic growth and development. Extensive technology shook down the ecological balance of nature. Deep mining, boring, extraction quarrying and blasting became the order of the day, which not only devastated the nature but displaced and evacuated rural and countryside masses who had nowhere to go. Technology induced modernization and industrialization, as we know happened much quicker than the social adaptation but it shook the foundations of social life everywhere (Caidens 1971: 147). For example the unaccountable scientific advancement produced income which could not be evenly distributed due to the ill managed use of natural resource and extractions. The greatest price of this unregulated affluence is the cruel income disparity in society. In India, despite the 13-folds increase in industrial production, rural and urban poor have been multiplying in numbers (A. Singh, 2000: 13).

"Income growth is not always translated into poverty reduction. This depends upon the ability of governments to use natural resources in a justifiable manner." The misuse of power of destructive technology during and after second world war was very well-experienced after the bombing of Hiroshima and Nagasaki which not only led to destruction and deformation of lakes, mountains, landscapes humans and animals but also the erosion of numerable earth forms like caterpillars, butterflies, birds and fish who sustain bigger life forms. This erosion of ecosystem base line changed the composition of soil temperature, humidity and climate (Ibid). Which gradually resulted into the phenomenon of global warming.

Administration and Development Policy

Science and technology was misused by governments during and after second world war. It became most powerful weapon in the hands of big powers to oppress the small powers as a result, the latter also joined the race of acquiring destructive technology to maintain state power. Indiscriminate use of science and technology to enhance the productivity and economic growth of the state became the order of the day. There was little realization among decision makers that impact of hazardous chemicals, aerosols and nuclear radiation was all directed against the land, water, air, forest, bio-mass, animals and humans. Competitive and ivory-tower decision makers were oblivious of the consequences the third world nations and its poor inhabitants had to face in order to develop. In the name of efficiency and productivity six per cent of worlds population took control over 60 per cent of world resources (Rachael Carson 1962: 4).

If the world thinks of ridding itself of poverty and squalor, the decision makers of international organizations like UN, WB, IMF and WTO need to work within the framework of environment otherwise no amount of conferences on Earth, water and Environment protection can save our planet from irreparable damage. Some similar notions were aired at the first conference on environment which took place in Stockholm in 1972 which said that removal of poverty is an integral part of the goal of an environment strategy for the world. The Rio-conference on Environment and Development 1992 further emphasized on anti-poverty strategy for sustainable development. It is the government which designs developmental policies and plays key role in distribution of resources to various sectors of health, education, defence, industries, power, nuclear researches etc. However, it is the particularistic forces which influence the decision making which is eventually implemented at the administrative level. In the net result, it is the politico-administrative-business trio which decides who will get what and how much. As the political designs are approved by the administrators, great onus lies with the administration to secure the interest of the people.

Existent Policies and Environment

The present approach of decision making is sectoral and disintegrated. Since the beginning of the Ist five year plan. Industrial development and urbanization was the topmost priority of the Indian Government. Heavy Engineering and steel industries definitely contributed in raising the living standards and urbanization of a vast population of India but it also had its reverse effects on double the population and area. On the one hand it impoverished and displaced the local inhabitants of their land and forest and on the other it denuded, extracted and mercilessly exploited the rich forest minerals, ores and fossil fuels without any returns to the local people. Foundation of steel and Iron industries of Durgapur, Bhilai, Rourkela, Ranchi, Bokaro are cases in point. There is acute shortage of electricity and water in these areas, factory production is of no use to the local consumption. Indigenous people called Tribals have no access to modern life and education. All these industries incurred heavy looses, few of them faced closure and others are reduced to 1/3rd of the working strength from which it had started. Underground coal fields of Jharia, Dhanbad, Chaasnala are burning for last several years posing grave threats to life and property. Mica and Coalfields of Bihar have become hallow and present serious environmental hazards due to heavy and extensive mining and extraction.

The exploitation of resources at grassroots for industrialization, transportation, urbanization and irrigation have led to complete dehumanization of environment. Only 20 per cent of the affluent people are enjoying the exploits of Nature at the cost 80 per cent rural population who have been robbed of their natural and cultural habitat. So far environment administration has been toeing the line of developed countries and falling prey to their standard of consumption without having respectively high economic growth. In order to imitate their way of living third world countries got themselves trapped into huge debts. the total external debt of developing countries has multiplied thirteen fold in the last two decades from $ 100 billion in 1970 around $ 650 billion in 1980 to around $ 2000 billion in 1994 (the Human Development Report 1992: 18).

Unfortunately, the westernization of third world governments and societies has not only eaten out their natural resources but has created sharp polarization in society about its consumption pattern. The social divide of pro-project and anti-project groups is manifesting into social movements like Tehri Bachao, Narmada Bachao, Chipko Movement and the likes. Medha Patekar was beaten up at Gujarat peace meeting during post Godhara riots because she happened to be the crusadar against the construction of 100 metre high Dam at Narmada Sagar. The state actually survives upon these hastily undertaken policies of resource extraction to meet the developmental parameter set by the IMF and World Bank. "Grassroot conservation" is considered a luxury which the policy planners cannot afford. As a matter of fact policy burdens of modernization and urbanization are all transferred to grassroots, which becomes the industry for primary resource production.

Administrative policies are directly responsible for environmental degradation. Since the launch of the new economic policy in 1991, India is rapidly advancing into liberalization of industrial policies. Multinationals are being given red carpet welcome. The government is withdrawing its role and disinvesting from public corporations and undertakings. The transnational investors are taking undue advantage of administrative loopholes. Given the pressure of poverty, population growth and soft administration, environmental standards and degradation assessments are often concealed and compromised. With the spread of liberalization business houses in India had transnational commitments more pressing than national obligations. (A. Singh, 1993: 23)

Eco-friendly Administration

The major challenge for administration is to formulate and implement eco-friendly public policies. Centre for science and environment, a leading environmental NGO in India in their second citizens report had laid down a framework for environmental action by government in the early eighties but official inaction resulted into environment Vs. development debate. Environmental priorities are being lost into administrative apathy and indifference. Thanks to Environmental movements which are drawing public attention towards sustainability and self reliance. There is an urgent need to sensitise the people towards the drawbacks of technology oriented development which rendered, the confused citizenry unemployed and useless generation after generation. Agricultural and family production units which were environmentally and culturally viable were discarded by declaring them retrograde and backward. Culture has intrinsic relationship with environment. When a village is selected for a mega project, it displace villagers through petty cash compensations and ignores the impact of development revival of traditional and indigenous modes of production displacement upon the children and old members of the community, needless to say that psyche of children and women due to break up and loss of identity undergoes irreversible shock and they end up into violent mob, recluse, mafia, prostitution etc.

Nature and environment have been degraded mostly because of the cost-benefit approach of the administrative policies. Projects after projects have been launched ignoring the real need of society merely on the basis of its cash and commercial utility. It was only after the Rio Conference 1992 that the intrinsic relationship between local communities and environment could be recognised. It strictly specified that local communities or locals should be treated as custodians of their environment and natural resources. Although the Panchayati Act 1996 recognizes the habitat rights of native communities yet the continued eviction of villagers in the name of promoting tourism and creating wild life sanctuaries continue unabatedly. Recent eviction of 24 villages from Sheopur area in Kinno district in M.P. in an endeavour to save Lions and Tigers is a case in point. In order to understand the local problem centralized bureaucracy must be replaced with decentralized decision making by the communities (Naresh Sharma, 1995, 2.2-2.13)

If there is any inclination to save the environment from degradation and pollution, eco-system approach will have to be given precedence. It is also called the integrated multi-dimensional approach. It deals with social, cultural and heterogeneous environmental issues. Environmental administration is a long way forward to development administration of sixties.

Communities are the basic units of environmental administration. Therefore, unless we have rural administration working properly environment cannot be guarded. Consequently strengthening of Panchayati Raj and creation of such institutions which can work in harmony with peripheral and grassroot necessities should be the immediate concern of public administration. Gandhian philosophy of village Republics will have to be revived. Grassroot participation in decision making will have to be ensured. The basic reason behind the failure of community development scheme launched in the decade of 50s was its nature of highly centralized planning. The same happened with high yield varieties programmes and green revolution. Rural development or Integrated Rural Development Programme (IRDP) all suffered the same fate because they were imposed from above.

Defective understanding of flood, draughts, land, water and forest have given rise to environmental blunders like mega-irrigation projects and Big dam irrigation schemes. The traditional agriculture and tank irrigation system has been neglected due to conventional attitude of bureaucracy. Only gradually the administration is realising the need of strengthening peoples participation in reviving land and forest. Few institutional mechanism like Watershed Association and watershed committee are being created. The committee is the executive body that functions with the Gram Panchayats (A. Singh, 1999).

It is high time that the bureaucracy shared its environmental concerns with the local NGOs and PRIs. Environmental protection cannot become a reality unless the bureaucracy is serious about it. Especially in case of India unless some labour intensive environmental strategy is envisaged it is impossible to control growing unemployment. The International Labour Organisation Report (1989) clearly warded off the fears of the industrial advocates in these words, "environmental management, clean technologies and conservation activities are highly labour intensive and are capable of a potentially important source of job creation rather than being the so called no-growth policies."

A large sum of public money is spent every year in poverty alleviation and rehabilitation schemes. If all these money could be instead used to absorb local people at initial stages in their own cultural vocations using indigenous technology, environment can be saved from further degradation. This would also help avoid confrontation between the local communities and the developmental experts.

Conclusion

Scientists and administration both will have to ackowledge the fact that commoners have lots of practical sense about environment conservation. We have had enough of Hi-tech development as envisaged by Nehru Administration. We are also facing the consequences of marginalization of Gandhian approach of rural development. That is why, no amount of community development programme launched by Nehru in 1952 could compensate for the damages done to the Panchayati Raj structure of rural administration even till today. Replacement of indigenous technology by big machines alienated local people from the entire process of development. Industries were operated and imparted training only through foreign experts and only highly qualified technical hands were absorbed in the second rung. Haphazardly planned industrialization with its focus on large scale production over stepped the energy storage and ecological sustainability. With burgeoning population, rate of energy consumption is not only proving too expensive but rather catastrophic to all developmental needs (S. K. Kak. 1998: 108). Therefore, reduction of energy consumption is the first need for environmentally sound technology. Or else we face the green house effect warming up the atmosphere emitted from the energy intensive industry. So far, state and administration have been granting excessive funding for ecologically non-productive research. Some researches must also be directed towards developing indigenous technology. Developed countries make intensive efforts to dump obsolete technology in third world which offers a potential market for their worn out technological innovations (Sardar and Rossier 1977: 538)

These energy intensive technologies are mostly borrowed from the west under technological exchange programme or technical aid arrangements under the austere guidance of technocrats and bureaucrats. People's requirements get only secondary priority here. The highly mechanized and energy intensive technology of Green Revolution which brought the country out of food crisis had torn the ecological and social fabric of the country because of its emphasis on over production. As rightly observed by Vandana Shiva, it has led to the ecological breakdown of nature and political breakdown of society by producing scarcity, not abundance. Similarly emission of poisonous gases and wasteful discharge of big industries such as food processing, iron and steel, non-ferrous metals, automobiles, air-conditioners, refrigeration, foams and polythenes, chemicals, pesticides and electro power generation are posing serious threats to environmental cover of earth. The most fatal impact of big technology had been the gradual disappearance of traditional occupations such as potter, weaver and other crafts men in India.

The use of peoples technology or indigenous technologies would revive these dying crafts and occupation with the help of modern science and not only enable common man to improve his employment potential without ecological degradation and dependence on industries but also protect the environment in a substantial way. Bio gas plants and improved Bullock cart technology are far better alternatives for rural masses than hi-tech expensive source of energy and ensure a better quality of life for them.

Bio-gas plants can help save the jungles from being burnt as firewood and fuel. There are many superfluous products like pepsi and mineral water drinks being marketed into plastic bottles by MNCs which pollute our environment. It is well within the jurisdiction of the administration to ban such products and confiscate their licences. On the contrary it is the Transnational National Companies (TNCs) which insist on resource controls through intellectural property rights. Patenting of Neem, Haldi and Basmati are cases in point. Needless to say it is important for Administration to device mechanism to protect our environment and natural resources from Draconian policies of WTO and MNCs. Big projects should be assessed in public interest and Environmental movements should be given proper attention and marginalization of subalterns, Tribals and poors must stop. Unless there is some initiative or strategy to involve common people into environmental economy, sustainable development and conservation of environment will remain a distant dream.

REFERENCES

1. Raghubir Singh, Paryavaran Awam Vikas, IIPA, New Delhi, 1999.
2. D. B. Ohm and B. J. Hiley, *The Undivided Universe: An Ontological Interpretation of Quantum Theory*, London and New York, Rout ledge, 1993.
3. Seyyed Hossein Nasr, *Religion and the Order of Nature*, Oxford University Press, New York, 1996.
4. Caidens, Gerald E. (1971). *The Dynamics of Public Administration: Guidelines to Current Transformations in Theory and Practice*. Drydin Press, Illinois, USA.
5. Amita Singh, *The Politics of Environment Administration*, Galgotia Pub. New Delhi, 1999.

6. Rachael Carson, Silent Springs, Hamish Hamilton, London, 1962.

7. Human Development Report (1991, 1992) UNDP, Oxford University Press New York.

8. Naresh Sharma, "*A Thematic Introduction to Local Markets in Traditional Technologies*", in 2nd Congress on Traditional Sciences and Technologies of India, Dec. 1995, Anna University, Chennai.

9. S. K. Kak, "*Energy Intensive Life Style Vs. Lifestyle for Conservation*", Book of Extended Abstracts, Third Congress on Traditional Sciences and Technologies, New Delhi, 1998.

10. Sardar, Z. Darwud Rossier Owen, "*Science Policy in Developing Countries*", in I. Spiegel-Rosing and D. Desolla Price (ed.), Science, Technology, Society, Siege, London, 1977.

11

Bureaucratic Politics in Environmental Control

Arun Kumar Upadhyaya*

Bureaucracy is considered as one of the most powerful instruments of Governance. It is through Bureaucracy that public policies are given shape as well as get implemented. Logically, therefore it has a significant role in the Governing process.

One of the biggest challenges before the State and its Bureaucracy today is the issue of Environmental degradation. Pollution has reached such a level that if urgent steps are not taken the very survival of life would be at stake. Such an alarming state of pollution is the outcome of our Technological advancement. Technology thus has alienated the Modern Man from Nature. At the same time Bureaucracy has alienated man from man. Both Environmental Pollution and Bureaucracy are an extension of the phenomenon called modernisation.

Bureaucracy existed even in the earlier known Empires of China, Russia, and the Great Ottoman Empire but the characteristics of present day Bureaucracy are peculiarly Modern; Hierarchical relations, salaried employees, standardised rules of procedures, impersonal authority, specialisation and dependence on written files for continuity. Modern Bureaucracy has come in for scathing attack by the people at large mainly because of its insensitive nature.

* Reader, Political Science, Government P.G. College, Chandauli, Varanasi.

Ironically steps to check environmental degradation call for bureaucratic action by which the drafting of the plan and implementation of the control takes place. Environmental movements can thus reduce pollution but rarely bureaucracy. But they can certainly make bureaucracy more accountable and less pre-occupied with its institutional interests. This would be not reducing the bureaucracy but reforming it alongwith its expansion.

New bureaucratic agencies have been created where environmental conditions have achieved issue status. Poland, Malyasia, Brazil, United States and Britain to name a few. The new agency is sometimes merely a token gesture and offers no assurance that the issue will be tackled. At other places the new agency is put under an existing ministry where it experiences uneven competition for funds and authority. In only a few cases the environmental agency is lucky enough to have autonomy, budget and political support to implement current regulations effectively and to take the lead in proposing legislative measures for itself. This shows that a new bureaucratic agency can itself be the victim of bureaucratization.

Sweden is a classic example of this irony of choices between the two brands of modern alienation. As early as the '70s the Swedes moved the curb environmental abuse and at the same time tried to dismantle their pervasive, centralised and elitist bureaucracy. To curtail potential and administrative arbitrariness the documents were required to the publicised and the institution of ombudsman established to handle public complaints against bureaucrats are two of these steps.

The Riksteig (Parliament) elections of 1973 were dominated by the twin issues mentioned above. The Centre party ran on a platform of nostalgia for the pre-industrial countryside and yearning for the non bureaucratic life. Party leader Thorbjorn Falldin exclude the spirit of a simpler Swedish Past when the streams were cleaner and relationship less highly organised. In reality Falldir and his party took up specific issues dominating the Sweedish *present,* tougher environmental regulations and bureaucratic de-centralisation. The Swedes who are essentially a conservative people have a big support to the Centre Party which polled 25.1 per cent of the votes and thus gave a warning to the winners, the social democrats to take their environmental and bureaucratic discontents seriously and to respond with genuine efforts.

This concern for the environment intertwined with a hostility towards bureaucracy often makes the environmentalists politically blind. The intricate working of the bureaucracy that affects environmental policies is ignored and the politics of legislatures and election is in focus. But in most effective environmental activists are those who carefully analyse bureaucratic politics and provide support to potential allies within bureaucratic departments.

Bureaucracy has been known for its twin sins of inaccessibility and unaccountability. Bureaucratic attributes have been its professional expertise and institutional continuity not affected by the uncertainties of popular politics. But citizens view even these attributes as shortcomings because they expect the Governments to not only competent and stable but responsible too. Environmental Agencies created in response to public demand also have the danger of following a prey to the sins that afflict the older organisations. Therefore the environmental activists have two contradictory tasks which regard to bureaucracy. A moral and often technical support to infant environmental agencies struggling for funds and intra-governmental cooperation and also constant monitoring of those agencies to ensure that they remain committed to environmental protection and accessible to the general public.

Institutionalisation and Environmental Agencies

Industrialisation refers to the process through which an organisation gains legitimacy and become agile and resourceful enough to survive changes of regime or policy. It is also a process by which an organisation achieves a routineness so that persons within it relate to one another in regular patterns and not according to personal likes or dislikes. Most environmental agencies are less than a decade old and not yet fully institutionalised. Some may never achieve, institutionalisation. Some fail to win confidence and cannot cope with political change; their mandates remain too vague and budgets too marginal for them to develop regular procedures. Eventually such agencies may be dismantled or simply shunted aside and denied any meaningful role in policy formulation or implementation.

A new environmental agency faces the problem of rivalry with older bodies that possess authority in areas or environmental importance. Such established departments like that of forestry, mines, agriculture, commerce, are more firmly institutionalised and have the resources and political contacts to ward off intrusions from any bureaucratic up-start. Because of the diffuseness of environmental problems a new regulatory agency is likely to be cast in the role of a governmental 'pocher' a role that undermines its efforts to gain legitimacy and cooperation.

The new environmental administrator has a wider constituency. He has "all the citizens" as his constituents, since environmental disruption threatens the well being of every one. Those environmental organisations that support and monitor newly created environmental agencies give the agency a more concrete and "bureaucratically" usable constituency as this would not make it "under privileged" due to lack of specific clientele interest and at the same time not captive of one client which robs it of its integrity.

The Impact of Origin and the Questions of Efficacy

An environmental agency may be handicapped from the outset because of the political and existing bureaucratic culture. For example one can quote Italy's bureaucracy which has been characterised as in efficient baffling and nightmarish. The national bureaucracy includes some one million people, most of whom are recruited by letters of recommendation from politicians or from among citizens afflicted by natural or man made disasters. After a 6 month trial they do not have to work unless they want to. Nobody on the state pay roll can be fired unless jailed for a criminal offence. In this situation compounded by instable party regimes in Italy it is not surprising that several million dollar international grant meant for the protection of Venice a city endangered by water supervision and polluted air were lost by the Italian Government. Italy even has a category of state book-keeping known as "residue passivi" which means 'money allocated for public investment but never spent because all trace is hopelessly lost in some ministerial bog'. An estimated 15 billion had been lost in this fashion according to 1973 figures.

One feature of an agency's origin that may prove important specially when it treads on the territory of pre-existing departments is its political sponsorship. The French ministry of environment proved effective because it owed its origin to a degree in 1971 by President Pompidou who later selected an energetic director who has a fellow Gaullist and party member. Similarly in Poland the ministry of regional planning and environmental control made a headway in its task because it grew out of the blue ribbon commission on the environment, in the affairs of which communist party leader and Prime Minister Edward Gierek was personally interested.

The person appointed to serve as the first director of a new agency can have a lasting impact on it. Maurice Strong, a Canadian was the first director of United Nations Environment programme and he had the advantage of having worked directly with the 130 separate nations that were to be serviced and advised by U.N.E.P. Strong utilised this personal prestige in an early conflict with the parent bureaucracy, the U.N. Secretariat headed by Secretary General Kart Waldheim Strong opposed cuts in the U.N.E.P.'s budget and managed to take the headquarter of the organisation from New York to Nairobi in Kenya much to the unwillingness of none the else but Secretary General Kurt Waldheim.

Resignation by a prominent bureaucrat may have to be not only threatened but also carried out in order to sustain the new bureaucracy. In West Germany, in 1972 Dr. Bernard Grzimek resigned as head of the federal governments 3 year old wild life protection commission. He objected to the light fashion in which ecology issues were being treated by the Prime Minister office, low budgetary allocation, lack of cooperation from other federal ministries and several other issues. Though his resignation was accepted but it led to an appreciable increase in the budget of the nature preserves and wild life protection.

Bureaucratic Mandate

The existing political culture, the characters the initial support from politicians, the leverage and independent stature of the first administrator and the adequacy of the first budget—all determine the future effectiveness of an environmental agency. To add to it is the legal mandate which is most fundamental because it defines its mission as well as authority. An original mandate can lend to three types of weaknesses. Firstly, the agency may be given two board a mission with insufficient resources which may lead to its loss of credibility and being reduced to being a bureaucratic "paper tiger". The American environmental protection agency (E.P.A.) suffered such a danger and had to face angry public protests in implementing clean air act of 1970. It had to seek the extension of the deadline while modifying or cutting most of its proposals. Secondly there may be a defective mandate which narrowly defines the mission of the agency ultimately causing it to have a little meaningful impact. The narrow mission may be compounded by the presence of modest staff and authority. This is what happened exactly to the high level agency created in 1973 in the Soviet Union to monitor air and water pollution nationwide. The new bureau was given no powers to resolve contradictory directives from other parts of the bureaucracy and/or the conflicting demands for growth as well as environmental protection coming from Soviet leadership. The agency protection coming from Soviet leadership. The agency was at pains to note that it could do nothing to stop the ministry of forest from becoming an exploiter than a protector as was originally intended. The ministry was allowing more and more trees to be cut in pursuance of the demands of the ministry of lumber and wood processing industry which was more concern with the bureaucratic targets of lumber production than the negative impact of soil erosion.

Lastly a mandate may be so ambiguous as to paralyse the new agency. An ambiguous mandate in the hands of politically talented and bureaucratically energetic administrator can produce a more wide ranging programme than the bills authors imagined. But that is more an exception than a rule and mostly this ambiguity makes a bureau vulnerable to outside criticism and plagues it with internal caution. Observers have found a mixture of both reactions to ambiguity in the American political and bureaucratic response to the national environmental policy act of 1969. While a federal Judge found the acts meaning "more uncertain" the very vagueness permitted the interest groups monitoring the E.P.A's bureaucratic performances to push the agency further.

Thus altering bureaucratic behaviour may call for amending the legislative mandate or undoing knots of interests that designed a weak agency in the first place.

Concept of Role

A role is a pattern of behaviour that others expect of any person in a certain position whether that position is Prime Minister, Civil servant, Forest ranger or a wife. To change a role is to change not personnel but popular expectations. Some roles may be well defined but others may be quite vague or controversial; Expectations about what is needed or what is legitimate in a post or in a flux or are contrary to one another. This poses the question, should the environmental administrator for example be an activist lobbying in parliament for tougher controls or should he be content implementing what is in the books?

Pressure to adapt one's behaviour to the expectation residing in a role is what is described by the formula "where you stand depends on where you sit". The formula suggestions that most officials and politicians are hypocritical and will adopt whatever policy stance makes their institutional life most comfortable. This may even change from one institution to another. But such alterations may not be dramatic. Some persons have firm personal ideas about public goals and private styles of action that move with them as they progress from one role to another. Role perceptions take a strong hand in shaping officials notion of their own self interest.

Role perception can be changed by several means most require not only an awareness of the complexity of role dynamics but also sufficient political influence to affect the change.

Another strategy for alterting role definitions is to bring a new public sector to bear on the agency. Ralph Nadar in U.S.A. adopted this tactic in his campaigns to make the American bureaucracies more responsive to the environmental constituency. In Italy perhaps the most effective environmental interest group working in this way is the large and prestigious Italia Nostra.

Bureaucratic Schizophrenia

Bureaucracies are as rife with conflicts of interests as any assortment of lobbyists or parties. Role conflicts in this context occurs when a single agency is saddled with competing or incompatible goals. One soviet commentator has described this situation as "asking the goat to mind the cabbage patch."

The American system has fewer instances of a government agency being simultaneously both industrial producer and regulator. But environmentalist have pointed to the Atomic energy commission as an example of such a role conflict.

Since role conflict undermines an agency's credibility a director would certainly deny that any ambiguity exists and assert the compatibility of the agency's responsibility. A.E.C. Director Dr. Dixie Lee Rayin the same vein denied that the A.E.C's promotion of nuclear technology compromised its role as regulator of nuclear power. Such role conflicts can be resolve through assigning priorities whether be it in individuals or bureaucratic organisations. There is another solution but this requires more profound political change than is involved in manipulating organisational charts and that is resolving the conflicts of public priorities which create role conflict. This might require public choices that would involve ideological and cultural alterations rather than mere bureaucratic reforms.

Bureaucratic Competitions

Bureaucrats working in different agencies perceive issues and goals differently. How that intra bureaucratic conflict is handled varies from political system to political system.

Since environmental agencies are late comers they are frequently seen by pre-existing agencies as threats or competitors. A new environmental ministry or department is a new competitor for budgetary allotments and for presidential or cabinet attention. If the agency is launched at a time of tightened governmental expenditure then the budgetary dimension of bureaucratic conflict is apt to be specially important. If on the other hand the agency makes it debut in an era of expanding governmental revenues and public expenditures, then the conflict with pre-existing bureaucracies may be muted.

Bureaucratic Clienteles

Inter-agency conflict is caused in part by the fact that officials in various departments see themselves as servants of special sectors of the nation. Agency identification with a certain group is the product of historic evolution—what groups needs prompted the government to create the agency in the first place. Identity also results from recruiting practices in which an agency relies on one sector of the public for its personnel because that sector has the training and interest that attracts them to the agency. Identity can be shaped as well by experiences in political conflicts during which one sector of the public has consistently provided support for the agency when it was under attack from other quarters.

The dynamic of clientele-bureaucracy relationship is commonly described as "pressure". This dynamics is often one of mutual support and re-enforcement. Therefore all interactions between administrators and interest group spokesmen are not adversary relations.

This type of interactions may turn into a "complex" relationship which refers to a recurring pattern of mutually rewarding relationships between certain government agencies and their primary non-governmental clienteles. Outside the United States the idea of a complex or sub-government is seen in the light of the degree of close interaction between government and private sectors presumed legitimate and natural.

In the environmental area several investigators have asserted that bureaucrat-politician interest group complexes do exist in the United States. Though the "complex" is an extreme form of bureau-clientele relations it reflects the more routine effort of administrators to cultivate sympathisers.

Conclusion

A cursory investigation of bureaucratic politics surrounding environmental policy is enough to reveal the prevalence of conflict within the bureaucracy and between bureaucrats and outsiders. Such conflicts appear to frustrate efficient environmental planning and regulations. It encourages permanent adhocism and the sacrifice of policy in the name of strategy.

Bureaucratic conflicts may be minimized by following certain conditions. Few of them are as given below:

1. A personally ambitious politician should not be appointed as head of a bureaucratic agency;
2. There should be an explicit and uniform ideological commitment on part of career civil servants;

3. Bureaucrats should have common educational and social class background irrespective of their technical expertise or professional allegiance;
4. The political party controlling the cabinet and the legislature should have permeated the institutional structure as well;
5. The legislative judicial or executive bodies that monitor the bureaucratic agencies should be relatively united and homogeneous;
6. There should be organisational techniques which are sophisticated enough to provide for adequate means to ensure inter-agency communication and co-ordination;
7. The society and economy regulated by the various agencies should not be marked by serious cleavages;
8. The revenues taken in by the government should grow as rapidly as the expenditure demands so that most agencies feel confident of receiving the budgetary allotments they need to sustain their own programmes.

Attainment of the above mentioned status would be a bureaucratic utopia but even then the environmental agencies have the chance of remaining weaklings. The harmonising conditions could serve the down play the governments responsibility for treating or preventing pollution. Therefore it can be safely concluded by saying that sometimes then the very existence of the conditions for bureaucratic conflict, give the environmental agency a fighting chance but at the cost of delay, waste and distrust in governmental operation.

REFERENCES

1. The Politics of Pollution in a Comparative Perspective; Ecology and Power in fourNations. Cynthia H. Enloe, Clarke University, 1978, Longman, New York and London, 1978.
2. Martin Schiff "Welfare State Bureaucracy and Democratic Control in Sweeden and Its Implications for the United States" (Paper Presented at the North Eastern Political Science Association, Buck Hill November, 1973).

3. New York Times, 19th September 1973.

4. "Sweeden", Which Way Now?" Special Supplement in the Economist 13th April 1974.

5. Linton K. Caldwell, "Environmental Policy and Public Administration", Policy Studies Journal-1, No. 4 (Summer 1973); 211.

6. Claire Sterling, "Italy," Atlantic, October 1973, p. 16.

7. William Murray "Letter from Rome", New Yorker, 8th April 1974, New York Times, 4th June, 1974.

8. William Murray "Letter from Naples", New Yorker, 26 August 1974, p. 55.

9. "How France Attacks Its Environmental Problems" World, 13th February 1973, p. 41.

10. Dan Morgan, "An Engls Vision Blurred by East Europe Pollution", Washington Post, 21st February 1971.

11. "Environment", Polish Perspectives (Warsaw), 16, No. 9 (September 1973); 63-64.

12. "Poland Giving Higher Priority to Environment", Environment News (E.P.A., New England Region), January 1973, p. 2.

13. "At the U.N.— A Showdown and Two Important Decisions", World, 30th January 1973, p. 69.

14. "The E.P.A.'s Contribution to Better Cities", Business Week, 4th August 1973, p. 21.

15. Leslie L. Ross & Richard A. Lixroff, "Political Dynamics and Environmental Policy", Policy Studies Journal 1, No. 4 (Summer 1973) 226-27.

16. Charles Conconi, "An Interview with Dixie Lee Ray", Environmental Quality July 1973, p. 21.

17. Robert Page, "The Image of the North's, Canadian Forum, June-July, 1973, p. 8.

18. Fritz Vilmar, "The Military-industrial Complex in West Germany and Consequences for Peace Policy", Journal for Peace Research, No. 3 (1973), 251-258.

19. Charles O. Jones, "The Limits of Public Support; Air Pollution Agency Development", Public Administration Review, 32, No. 5 (September-October 1972), p. 502-08.

20. John C. Esposito, "Vanishing Air; Nader Report on Air Pollution (New York Crossman Publishers, 1970).

21. Ian Mcdougall, "The N.E.B.; Let's Give Credit Where Credit in Due", Canadian Forum, June-July, 1973, p. 30-31.

12

Environment and Administrative Role

Dr. Rajendra Bharati*

Survival is the basic instinct which is found in every living organism and struggle for survival is practised by all organism in all the time and all the ages. But this can be possible when the milieu in which they are living can't be altered or threatened by foreign elements. The foreign elements are in the form of pollution both physical and social. This survival instinct has both positive and negative manifestation. When it is in form of positive thought, lead to conservation of environment and development in true sense, but when it is in form of negative thought, lead to destruction of environment, which not only posed problems for present generation but also for future generation. In the late 20th century problems emerged because of environmental pollution compelled people to think not about its conservation but preservation too. In fact conservation of the environment is the foremost need of our world today.

Environment and Pollution

Air, water and land comprise our environment. In geographical term we call it atmosphere, hydrosphere and lithosphere respectively and all these constitute biosphere. This biosphere consist not only of plant, animals, birds, fishes, insects but micro organism like algae, fungi, bacteria also. Whenever a change,

* Senior Lecturer, Department of Political Science, Simdega College, Ranchi University.

physical or chemical, occurs in the atmosphere and/or hydrosphere as well as the lithosphere all living beings get affected. This change is termed as pollution and the agents that institute these changes are called pollutants.[1]

The concern of environment is not only related with physical environment i.e. lithosphere, hydrosphere and atmosphere but also with social and humane environment i.e. social political, economical and cultural environment. It is proved fact that both the environments affects and being affected by each other.

A man can live without food few days without water one or two days but it is impossible to live without air for few minutes. Air is the mixture of gases comprises Nitrogen (78%) Oxygen (21%), Carbon dioxide (0.03%) and less than 1 per cent other gases like organ etc. Whenever balance in mixture of gases disturbed, air pollution is caused. The main agent of air pollution are dust from quarries, fume and dust from thermal power, gases released by the automobiles etc. The air pollution affect human health in the form of respiratory diseases ranging from common cold to lung cancer.

Land pollution is caused by using chemical fertilizers and pesticides like D. D. T. This has been used to increase agricultural production. But in fact they not only reduce land fertility but also affecting acqua culture. The presence of pesticidal residues in vegetables, milk, breast milk, fish and meat has been affecting the human health.

Water is the blood streams for every plants and animals. The pollution of water poses problems not only to human beings but plant and animals too. The water pollution is caused by chemical fertilizes and effluent drained into the river or dump into grounds which after precipitation mixed with ground water. Another pollution is noise pollution which is caused by industrial noise. transport, use of high noising equipment. Because of noise pollution hearing capacity of human being reduced which ultimately affecting brain and nervous system.

The main cause of ecological dis-balance is due to depletion of forest. In fact forest is considered as "green gold" of a country.[2] It play a vital role towards sustaining our life and national economy.

They not only provide gaseous balance in atmosphere but also a very important moderator of the climate. They not only control flood but also control soil erosion and landslide. The rate of depletion of forest in India is very fast due to following causes:[3]

1. People feel that forest are open treasure houses which they can use as they wish to use to fight against ever increasing population and poverty;
2. The greed of forest contractors and nexus with official axe both new and old trees for commercial purpose;
3. Mining is also a major cause of denudation of forests;
4. The conversion of forest area into cultivable land. Apple and tea cultivation are prominent agricultural practices in converted forest areas.

Pollution of any type not only poses problems for physical environment but also social and human environment. Since pollution affect human health and which affect social cohesion and ultimately left effect on political, economical and cultural health of the society in long run.

Administration and Bureaucracy

The term 'Administration' literally mean "to manage affair", "to look after people" etc. But it has been seen as part and parcel of civil society which is moving and strive for progress and development. In fact administration is a moral act and administrator is moral agent.[4] Need of effective administration is felt more in the countries like ours because it has to face the challenges posed by national and international affairs. The administrative organization is the only effective tool which could be able to tackle the problem so arisen. Administration here refer to mobilization of resources, human and material to achieve pre-set objectives.[5] It is an action oriented organization which can be able to fulfill the desire of both people and government. It has a planned institutional capacity to accomplish the specific goals of development through the formulation of appropriate policies, programmes and projects and their successful implementation. The essence of such administration is participative, responsive and accountable management.[6]

The bureaucracy is the arm of administration or in other word they are the mean through which goal of administration or government can be achieved. The term bureaucracy being generally used in derogatory term as senile, dormant or reactionary forces. But in its positive term it plays a vital role to sustain administration or government or society at large. Max Weber considers bureaucracy as "rational rule abiding" organization. The need and effectiveness of bureaucracy is felt more because they have direct relation with the people as well as public figures. They are very effective in making, implementing and evaluating the plans.

B. Guy Peters[7] has discussed the resources of bureaucracy in following heads:

1. Information and Expertise

The first and perhaps the most important resource of the bureaucracy is information and expertise. Information has been described as power and this power is always with the bureaucrats in their files. They have most of the data which is very essential in formulation of policy.

Robert K. Merton has rightly said, "The chief merit of bureaucracy is its technical efficiency with premium place on precision, speed, expert control, continuity decision and optimal return on inputs."[8]

2. Power of Decision

Despite red tapism administrators are in better position to make efficient decision. Bureaucracy have few procedural rules for free discussion but they can take decision earlier than any other institutions like political parties or legislatures.

3. Political Support

The administration uses to contact with various pressure and interest groups. These groups are formed on certain political ideologies and interests. In fact these groups have their own political and social link. By their support administrator can be helpful to get support for their plans and policies.

4. The Advantage of Being Apolitical

Since the administrators are appointed on non-partisan character and along with this they have expertise in their field which lead them to have a greater say in decision making. In addition to this, they can make a strong argument that their decisions are superior on technical grounds to those that might have been taken by political institutions.

5. Development of Agency Ideology

The bureaucracy develop agency ideologies as means of justifying their actions. These ideologies are important weapons in the struggle for influence. Outside "non-expert' feel helpless before the self serving agency.[9]

B. Guy Peters has rightly remarked that the existence of such an ideology is important for success of the agency in dealing with political institution.

Myth of Development

Development in third world countries are considered as to copy western way of life in all sphere political, economical and social. Development literally mean developing or being developed in all senses. In other way it is related to the area where new industries are established for achieving employment, goods of livelihood etc. But this concept of thought emerged from western materialistic society where ideas are flourishing to achieve maximum luxury by exploiting natural resources. This concept of west has been flourishing in third world countries also; on the assumption that western style of progress is possible for all. But Rosaluxemberg has pointed out that early western capitalism was achieved due to colonization of third world countries from where they got natural resources at cheaper rate and sold them after processing on higher rate and thereby also damaged their ecology and by making disbalance in biodiversity system. In fact western concept of development was based on continuation of the process of colonization whether physical, economical, social or ecological.

'Development' could not but entail destruction for women, nature and subjugated cultures, which is why throughout the third world women, peasants and tribals are struggling for liberation from 'development' just as they earlier struggled for liberation from colonism.[10] This paradigm of development is based upon the concept which consider all works that do not produce profit and capital are unproductive one. This concept of development sees only present day's luxury not foreseeing future which would be bleak and dark. In fact exploitation of natural resources ultimately endanger the life of present day also by damaging environment and disbalancing biodiversity system.

When commodity production as prime economic activity is introduced as development, it destroys potential of nature to produce life and goods and service for basic needs. More commodities and more cash mean less life in nature (through ecological destruction) and in society (through denial of basic needs). It is no accident that the modern, efficient and productive technologies created within the context of growth in market economy terms are associated with heavy ecological costs. The resource and energy intensive production processes give rise to demand over increasing resource withdrawals from eco system. These withdrawal disrupt essential ecological processes and convert renewable resources to non renewable ones. A forest for example provides inexhaustible suplies of divers biomass over time if its capital stock is maintained and it is harvested on a sustained yield basis. The heavy and uncontrolled demand for industrial and commercial wood, however, requires the continuous overfelling of trees which exceeds the regenerative capacity of forest ecosystem and eventually converts the forest into non renewable resources... . Sometime the damage to nature's intrinsic regenerating capacity is impaired not by over exploitation of a particular resource but, indirectly, by damaged caused to other related natural resources through ecological processes. Thus the excessive overfelling of trees in the catchment areas of streams and rivers destroys not only forest resources, but also renewable supplies of water, through hydrological destabilization. Resource intensive industries disrupt essential ecological processes not only by their excessive demands for raw material, but by their pollution of air, water and soil.[11]

In the era of globalization and liberalization the organizing principle for natural resources use is the maximization of profit and capital accumulation. Nature and human needs are managed through market mechanism. The ideology of development in based on largely to bring all natural resources to market for commodity production.

It is a fact that each and every society strive for development and progress. There is a need of a mechanism and organization which would be able to fulfill the desires and needs of the people at large. It is impossible to deny the concept of development, but the need of the hour is to redirect the process of development on long-term basis in which ecology can be sustained. There is no alternative to the development but an alternative is there for western model of development in which people's participation and biodiversity balance can be achieved. The main thrust of development is to preserve human dignity and social cohesion.

Environment and Its Management

Management of both the environment i.e. physical and social is the basic thrust area of administration these days. As we know that necessity is the mother of invention and new things are invented either due to need of hour or the exercise of mental faculty of man who is rational being and always tries to go deep into happening surrounding him. In that way we come to accost with technology which is the foundation of any industrial development.

Technological revolution gives boast to economic development. But this economic development has to be in direction of sustainable development but the truth is otherwise. Sustainable development aims to balancing the current needs of life vis-a-vis future need through resource conservation, focusing on minimal or no adverse impact on total environment.[12] Sustainable development leads to happier, tension free and humane social which also includes management of environment.

Traditionally environment management is related to protect physical environment only but this one is a myopic view. Management of environment need to have a wider overview. It should include political, social, economical, technological environment along with physical environment. Some time pressure exerted on administration by the society and people to manage total environment. Manifestation of such upsurge compel the executive, the legislature and the judiciary to act in such a manner so that right to life can be possible to all.

Constitutional Provisions and Judicial Pronouncements

Art, 21 of the constitution of India clearly says that "No person shall be deprived of his life or personal liberty except according to procedure established by law."

Though the article appears to be negative in its grammatical form, it has, in reality, been given a positive effect by judicial interpretation.

Life in Art. 21 is not merely the physical act of breathing. This has been recognized by the courts. In fact as philosophers tell us, life is lived at many levels. The *Rig Veda* (10.177.2), gives subtle description of the mundane activity of speech ... while the external mundane activities of life have their own place. They are manifestations of an inner, unseen, unperceived activity - which indeed a real life, that a human being lives. It is true that judicial decisions on Art. 21 do not embark upon such an analysis in depth. But they took note of the width of the right to life.[13]

Supreme Court in one of its judgement declares right to life include right to healthy environment. [M.C. Mehta Vs. Union of India A.I.R. 1987 SC 1086, 1090; Bandhua Mukti Morcha Vs. Union of India AIR 1989 SC 802] Supreme Court also declares that right to life under Art. 21 includes the right to enjoyment of pollution free water [B. L. Wadhera Vs. Union of India. AIR 1996 SC 2969].

In the case of M.C Mehta Vs. Union of India 1997 3 SCC 715. Supreme Court declares that the "Precautionary principle" requires the state to anticipate, prevent and attack the causes of environment degradation.

In a separate case of Vellore Citizen Welfare fForum Vs. Union of India AIR. 1996, SC 2715, Supreme court declares that chemical or other hazardous industries which are essential for economic development may have to be set up. But measures should be taken to reduce the risk of hazard or risk to community by taking all necessary steps for locating such industries in a manner that would pose least risk or danger to community and for maximizing safety requirements in such industries. The Supreme Court also direct the High Court to set up a "Green Bench".

Art. 48 A of constitution of India states that the state shall endeavours to protect and improve the environment and safeguard the forest and wild life of the country.

Administrative Role

Administration these days become part and parcel of every civil society. It is impossible to think any system without administration. In fact decay of civilization is due to decay of administration. Thus the role administration is very essential for management of total environment. its role can be seen in following lines:

(i) To educate people about environment protection and the dangers that emerge in future from degradation of environment;

(ii) To enact laws and judicial pronouncements in true spirit;

(iii) To provide help to the NGOs who are working in field of environment protection and conservation;

(iv) By providing inputs to the legislatures so that they (legislature) can be able to prepare a policy for development which would be eco-friendly;

(v) To arrange workshops in villages and towns about environment protection;

(vi) To utilize women force towards environment protection because both have the natural regenerative capacity;

(vii) To make strict vigilance over protected areas by involving local people;

(viii) To encourage people's participation in the management of environment;

(ix) Administrators can be helpful to educate people about benefit of family planning which ultimately put less pressure on exploitation of natural resources;

(x) Administrators should involve local people towards recycling of water, forest etc. by using conventional methods like check dams in plateau areas of Jharkhand;

(xi) Administration can be helpful towards establishing small hydelpower generating plants which can be helpful in reducing air pollution and tackling the problems of displacement of large inhabitants of the areas, where such plants being established.

In nutshell role of administration is educative, participative, accommodative and innovative in field of management of environment as a whole.

Conclusion

Every experiment and invention in all the branches of science is for the happier and luxurious life. But this happier life can only be achieved by making bridge between development and societal norms. Running away from the developmental concept would lead to isolationism which in long run convert into the problem of alienation, frustration and insecurity among the people. Thus the need of hour is to have a human face of development with active involvement of administration.

REFERENCES

1. Manivasakam, N., "We Breath and Drink Poison" NBT, New Delhi, 1995 p. 1.
2. ibid p. 87.
3. ibid p. 90.
4. Tead, Orway, "Administration: Its Purpose and Performance" New York, Harper & Brothers, 1959 p. 67
5. Maheswari, Sri Ram, "Administrative Theories" Allied Publishers Ltd. New Delhi 1994 p.1.
6. Battacharya, Mohit: "Public Administration: Structure, Process, Behaviour", The World Press Pvt. Ltd. Calcutta, 1987, p. 285.
7. Peter, Guy, B. "Copmpairing Public Bureaucracies: Problems of Theories and Methods". The University of Alabama Press, Tuscaloosa, 1988 p. 188-190.
8. Merton, Robert, K. "Bureaucratic Structure and Personality in Robert K Merton et al. (Eds) "Reader in Bureaucracy" The free Press Chicago, 1952, p. 363.
9. ibid 190.

10. Shiva, Vandana; "Enviornment for Women's Empowerment" in Kapur Promilla Ed.: "Empowering Women" Publication Division New Delhi, 2001, p. 185.

11. ibid, p. 189-90.

12. Khader S. A., Yojana February 2003, p. 20.

13. Bakshi, P. M., "The Constitution of India", Universal Law Publishing Co. Pvt. Ltd. Delhi, 2002. p. 48.

13

SAARC and Environment

Dr. Kalpana Chakravorty*

The concept of SAARC viz. South Asian Association for Regional Cooperation came to a forefront after the unprecedented success of ASEAN (Association of South East Asian Nations). The Far East nations realized the futility of aligning with either of the two powers and they rejected their hegemony and formed a consensus on the formation of an association which will work for the overall development of member nations and will be free from the opportunistic warmongers and exploiters in the cold war era.

The member Nations of SAARC (India, Pakistan, Bangladesh, Sri Lanka, Nepal, Bhutan and Maldives) had adequate history & culture to share; a prerequisite for the formation and success of an association or group. The member Nations, all being a part of British Commonwealth were equally footed in their aspirations & realization of goal of self reliance through interdependence. Besides all Nations being part of Indian subcontinent were part of a unified geographical entity and were all young Nations dealing with similar problems of poverty, unemployment, utilization of resources, energy security, strategic concurrence, American influence, challenges of free trade etc. Further British Commonwealth and its Colombo plan which preceded the SAARC paved the way for the member Nations to realize the goal of economic self reliance through co-operation. Since its birth, SAARC was considered as a 'cadaverous' organisation in comparison to successful groups like ASEAN and

* Ph.D. in Psychology, Senior Teacher, Sunbeam School, Varanasi.

others. The reasons were quite obvious the parallel seedy relationship of member countries like India and Pakistan, Maldives, Sri Lanka, Bhutan, in terms of their size, economies, human resource etc will definitely be considered as invidious, besides questions were raised regarding the success of forged co-operation, with the on going disputes between the member Nations.

Though Pt. Jawahar Lal Nehru & Indira Gandhi initiated steps for the forging of co-operation among nations of South Asia, the sole pioneer behind the formation of SAARC was the then Bangla President Zia-ur Rehaman. In 1981 he presented such reasonable proposals that both India and Pakistan accepted the document albeit the differences. In 1983 a meeting was convened of all the seven foreign ministers in which different deliberation were conducted and programmes were formulated for the 1st summit. The 1st summit level meeting of SAARC took place in Dhaka in 1985 in which the 14 points "Dhaka declaration" were unanimously accepted, after the signing of "Dhaka declaration" acceptance of IPA (Integrated action plan) by member Nations, SAARC formally came into being and started functioning.

The Heads of all States or Government at their first SAARC summit held at Dhaka on 6-7 December 1985, adopted the charter formally establishing SAARC. The following objectives can be laid down as adopted from the charter:

1. to promote the welfare of people of South Asia
2. to accelerate economic, social and cultural developments in the region.
3. to promote collective self reliance.
4. to strengthen co-operation among developing countries.

For the successful achievement of the defined objectives, SAARC has given recognition to a host of bodies that dwell in different fields various spheres of human development of the member countries. A host of recognized bodies are:

SAARCH—in the field of architecture

AMDISA—in the field of management

SAARC FUW—welfare of women

SAFA—accountancy

SAARC—association of town planners

SAARC—Cardiac society

Association of SAARC Speakers & Parliamentarians

Co-operation with International & Regional Organisations with UNCTAD—MOU on trade analysis & information system Feb. 1993 between SAARC & UNCTAD with UNICEF—Dec. 1993

(3) APT 'SAARC-APT

(4) with ESCAP

(5) with UNDP

(6) with UNDCP

(7) with EC

(8) with ITU

(9) with CIDA

SAARC & Environment—Ever since the industrial revolution in 19th century, the relation between man and his environment has been bedeviled by the economic activities undertaken by him. The greed of man in the utilization of natural resources has taken a toll of the environmental balance and has put it in a precarious situation. The globalization wave has augmented the exploitation of precious natural resources; the call of economic prosperity has further dented the already frail environment. The pace of industrialization is like never before and this has resulted in the cropping up of unlimited environmental problems, New environmental phenomena are being discovered like hypermelting of glaciers, unprecedented fall in temperature in the higher reaches of Dubai, disappearance of vultures from North India, presence of pesticide in underground water in Kerala, bleaching of Corals in Pacific ocean, disappearance of cranes from their normal migratory abode etc., likewise several unnatural phenomena can be mentioned which suggest a change in environmental homeostasis. SAARC as a responsible regional organisation has taken a grave concern over the environmental changes in the subcontinent; it has addressed these issues & has taken significant steps.

It was in 1987, after only two years of its birth SAARC decided for the establishment of a commission to study the causes and consequences of natural disasters & protection & preservation of environment and another study was undertaken about the concerned effects of the emission of greenhouse gases, which was ratified by the seventh summit in Dhaka 1993.

The summit of 1995 in Delhi called for the implementation of the recommendation of the previous two summits, accordingly the environmental ministers of member nations met and an immediate action plan was formulated, a plan of this respective was also presented at Kyoto conference which received widespread applaud.

The action taking committees have identified several areas and defined modalities for their successful implementation, which among various fields include better upgradation of tech-facilities, for better monitoring of climatic phenomena & integration of different sources through SAARC meteorological research centre, developing climatic change & sea level rise scenario through country specific studies & sharing of information data in this respect, making available to member states expertise climatic research & monitoring green house emission, exchange of information & data exchange of experience on strategies for developing , mitigating and adaptive responses to climatic changes.

SAARC & Regional Environment Under Different Summits— Basically SAARC is an economic association, working under the impulse of ASEAN, and its success as an economic and political organization, but issues like environment are not and cannot be relegated, SAARC too has taken the ramifications seriously and has taken a serious stance to cope over this problem. Environmental issues have found a prominent place in different summits and many times taken a centre stage, where political differences have always marred the progressive issues.

In the Third Summit at Kathmandu (Nepal), 4 Nov. 1987 the members of SAARC for the first time undertook the issue of environment and expressed their deep concern over the fast continuing degradation of the environment, the invasion of urbanization towards forest area & the debilitating forest cover. The analysis of frequency and amplitude of natural disasters affecting South Asia like cyclone, land slides, tidal waves were tabled and further deliberations were conducted. The green house effects & its consequent rise in sea level were taken into cognizance. The summit unanimously accepted that frequent natural disasters occurring in South Asia are a part of global change in environmental balance which needs to be taken seriously as these too inhibit the progress of development process and prospect of member Nations. They decided to commission a study for the protection & preservation of environment & to the cause and consequences of natural disasters in a well planned comprehensive network.

In the joint press release the members declared their intention to initiate the process of dealing with the problem of rapidly deteriorating environmental conditions in the subcontinent, they accepted the enormity of situation and pledged their all possible contribution to curb the further change in environment and possible natural disasters affecting the region. The secretary General was entrusted with the duty of continuous monitoring and proper implementation of the undertaken environmental programmes.

At the Fourth Summit at Islamabad (Pakistan), 31 Dec. 1988, the Heads of States or Government expressed their deep sense of sorrow and profound sympathy at the loss of valuable lives & extensive damage to property as a result of unprecedented flood, cyclones & earthquakes. The SAARC nations accepted the cascading effect of environmental deteriorations with respect to the growing frequency of natural disasters affecting the region. In this respect they revised their earlier decision taken in Kathmandu Summit (1987) to intensify regional co-operation with a view of strengthening disaster management and preservation of environment, they further took note of the recommendation of the SAARC group of experts, that met in Kathmandu July 1988, the joint press release envisaged that identification of programmes & measures as prescribed by a group of experts would supplement national, bilateral, regional and global efforts to deal with the increasing degradation of environment. They further urged for the immediate completion of research works related to the decline of different parameters, so that it could provide a basis for the member countries to draw upon action plan for meaningful co-operation amongst member States. Their joint statement further emphasized the concern over the change in global heat distribution due to widespread economic activity; a joint study was decided to be undertaken on the Green House effect and its impact upon the region.

In the Fifth Summit at Male (Maldives), 23 Nov. 1990, the Member Nations unanimously accepted the decision of council of ministers in regard to the time frame for the completion of regional studies on the causes of frequent natural disasters and protection of the environment. They satisfactorily accorded the methodology for undertaking the study on the 'Green House Effect' and its impact on the region. They hoped that the result of the study will be implemented by the member Nations at the Sixth summit. The SAARC member Nations recognized that the increasing depletion of rainforests all around the world was one of the major reasons for the change in normal environmental conditions and hence they introduced this new study in monitoring the global homeostasis. They expressed their hope that these studies will augment the efforts of an action plan for meaningful co-operation in the field of environment and disaster management.

The Heads of the States were shocked by the startling predictions made by the Inter governmental Panel on Climatic Changes (IPCC) which was constituted by the SAARC to calibrate the long term effects of global environmental change. A call for international support was made at the behest of SAARC, the member Nations affirmed that the change in global heat balance has threatened the existence of Island countries like Maldives, Bangladesh, which is to be protected at every cost. SAARC accepted the dearth of funds as a major hurdle in research and scientific development works especially for a developing country. The member Nations pleaded the international community to mobilize additional finance and to make available appropriate technologies to enable the developing countries to face new challenges arising from climatic changes and sea level rise. SAARC nations accepted the environmental deterioration as a major scourge to all mankind and resolved to help and co-ordinate every global effort towards environment. They also decided to celebrate 1992 as the SAARC Year of Environment.

On Nov. 23, 1990 at Male in the joint press release, SAARC made it clear that all the regional study on the causes and consequences of natural disasters and the protection and preservation of the environment and the study of green house effect be finalized before next summit. They suggested that before the submission of report member Nations should take certain preemptive measures to control the damage done by rigorous economic activity.

The Sixth Summit at Colombo (Sri Lanka), 21 Dec. 1991, upheld environment as one of the most debatable topics and SAARC nations recognized that the degradation of environment has emerged as a major global concern. The SAARC nations squarely indicted the apathetical attitude of developed countries in curbing environment harmful practices while it accepted that protection of environment is everyone's duty it exhorted that the biggest responsibility rests upon the developed nations as they are the biggest contributors to the global pollution by the use of environment unfriendly techniques and also they are well equipped technically and financially to undertake damage control measures.

SAARC too identified the new global culture of commercialism, which has revolutionized the life style of people. They called upon the international community to address the question of unsustainable production and consumption pattern which leads to a rapacious exploitation of natural resources. They also rebuffed the commensurate measure of environmental standards in both developed & developing countries as it would have excessive and unwarranted social cost if applied to developing countries.

The member SAARC Nations called for adequate new and additional resources that are needed for developing countries to pursue environment friendly techniques, without affecting social sectors. Such resources should be transferred to developing countries on an assured basis, along with environmental sound technology at preferential and concessional terms.

The SAARC member nations expressed their convictions that a frame work should be laid down for an equitable global partnership of all countries with a view to safe guard the environment.

After the completion of regional study on the causes and consequences of natural disaster and the protection and preservation of environment, the Heads of the States expressed their satisfaction and urged the member Nations to promote co-operation amongst themselves for enhancing their respective disaster management capabilities and for undertaking specific work programmes for the protection and preservation of environment. The green signal was given to the establishment of a committee on environment, in this respect the recommendations of council of ministers was accepted to

- Examine the recommendation of regional study.
- To suggest possible action.
- to decide on modalities and necessary feed backs for their successful implementation.

The Heads of States or Government expressed their satisfaction over the decision of standing committee, concerning time frame to finalize the regional study on the "Green House Effect and its Impact upon the Region." They urged the member Nations for greater co-operation on key issues concerning climatic change. It further articulated the new fields like Biodiversity, Biotechnology, and consensus on forestry principles, management of waste and other issues for discussion at Inter governmental negotiating committee and the UNCED in 1992.

In the Seventh Summit at Dhaka (Bangladesh), 11 April 1993, the member Nations of SAARC expressed their profound concern at the unabated degradation of environment. The Nations stressed the issues of environment as one of the most prominent of all economic & social issues, they reacted with anguish at the apathetical attitude of certain states & agencies in controlling this menace , it questioned their integrity and of the so called humanitarian issues for which they are fighting. The SAARC nations expressed their commitment to implement at all levels the programme for the preservation and protection of environment. The Heads of state government lauded the completion of the 'Green house effect', a regional study which was required for the availability of certain feedbacks before taking some decisive actions.

The Heads of States or Government applauded the developments of the United Nations Conference on Environment and Development held in Rio de Janeiro 1992 & highlighted the need to ensure that flow of resources to successfully implement the wide range of suggested measures contained in Agenda XI. The member Nations called for an integrated approach in taking actions and to consider the protection of environment, economic growth and eradication of poverty as mutually reinforcing. The environmental issues should not be used as a shield as by the developed countries in order to gain unfair trade advantage. All environmental initiatives must preclude the imposition of arbitrary and one sided pre-conditions and must ensure a harmonious blending of environmental and developmental concern. They further extended their support to the establishment of the commission on sustainable development; they urged the commission to take an unbiased approach of the environmental problems of developed and developing countries in an equitable manner.

The Heads of States or Government hailed the spirit shown by each member, which reflected the character of SAARC, and expressed their willingness to continue in future the practice of intra SAARC consultation on matters of common concern.

The Eighth Summit at New Delhi, 4 May1995, again gave environment a centrestage; the Heads of States or Government expressed their satisfaction over the formulation of a common position by member States before the World Conference on Protection of Environment at Yokohama Conference. They expressed their shock and awe at the ineffectiveness of certain damage control measures and the continuing degradation of environment. The member Nations expressed their commitment to forge coalition at every level national, bilateral, regional and global, programmes for the protection and preservation of environment. SAARC recognized the importance of international co-operation in building of national capabilities, transfer of ecofriendly techniques, research efforts in natural disaster reduction.

The Heads of States or Government accepted that actions related to environmental degradation have helped in the area of disseminating information of the environmental problem and its effect, yet in the food proof execution of the recommendation of the SAARC regional study on environment, causes and consequences of natural disaster and study on green house effect, which were completed by 1992, progress has been slow. It directed the agencies for the effective implementation of the recommendation of two studies, without further delay, it further directed the technical committee on environment to monitor and address the recommendation of previous studies and submit their report in the ninth summit through council of ministers. SAARC nations reiterated for the committed 0.7% of GNP as official development assistance by the developed countries as promised at the United Nations Conference on Environment and Development.

The Heads of States or Government also expressed their satisfaction at the functioning of the commission on sustainable development and expressed their hope that the commission would be able to facilitate the necessary flow of resources and technology. The SAARC nations expressed their happiness that the convention on biodiversity and the UN frame work convention on climatic change have come into force, welcoming these developments SAARC further stressed the need of greater assistance to developing nations, in terms of financial and technical help so that the process of economic self reliance can be easily integrated with the preservation of environment.

The Ninth Summit at Male (Maldives), 14 May1997, witnessed the SAARC Nations expressing their deep concern over the increasing pace of industrialization at the cost of environmental degradation; they conceded that the growing poverty and population in the developing countries and the rapid commercialization in developed countries are two of the biggest contributors to environmental pollution and degradation. It further called for the increasing flow of resources from developed countries for speedy functioning of the various ecofriendly agencies and world bodies. The Heads of States expressed urgent need for the speedy implementation of the recommendation of SAARC regional study on the green house effect. The member Nations further discussed the feasibility of a regional treaty on environment at a parallel to the international treaty, they further stressed the need of cooperation among members and developing scientific capability to protect and manage the environment. SAARC nations agreed on the necessity of controlling air and water pollution and for common minimum standards through exchange of information, strategies and technology. SAARC affirming the worth of bio-diversity in the south Asian region pledged its commitment towards the protective and conservation of biotic resources with especial interest towards the endangered species. They took note of the increasing illegal trafficking of flora and fauna from the region, and looked forward to the possibility of concluding a convention on it. They also called for the early execution of United Nations frame work convention on climate change and adoption of a protocol to limit the emission of green house gases by developed countries within an early time.

The SAARC nations emphasized the importance of an effective and unified position of the association, at the coming session of the UN general assembly to apprise the implementation of Agenda 21.

In the Tenth Summit at Colombo (Sri Lanka), 31 July 1998, the Heads of States once again called for the effective and early implementation of SAARC action plan, it further endorsed the initiative taken by Maldives for the establishment of a coastal zone management centre, it also ordered the member governments for the preparation of national environment action plans before the end of 1998.

The member Nations lauded the ratification of Kyoto Protocol to the UN frame work convention on climatic change in Dec. 1997; it emphasized the importance of the protocol for the protection and preservation of environment. It expressed its disappointment at the American rebuffing of Kyoto protocol on the commission of green house gases and urged all the nations to ratify it. The Heads of state government framed agenda for the meeting of SAARC environmental ministers to be held in Sri Lanka 1998, and asked the ministers to constitute specific measures required for SAARC to further strengthen cooperation on environment issues with member nations as well as with international organizations in the same field.

The member Nations in the Eleventh Summit at Kathmandu (Nepal), 2-3 Jan. 2002 expressed their satisfaction with the growing public awareness about the need for protecting the environment within the frame work of regional cooperation. They reiterated for the earliest possible implementation of SAARC environment plan of action, and directed the environmental ministers to take effective steps in this regard. SAARC nations strongly called for the need of cooperation in the field of early warning systems for the successful prediction of natural disaster and management of programmes related to conservation of land and water resources. The SAARC Nations also called for the establishment of regional bio-diversity database, in terms with the UN convention on Biological diversity and to develop a co-operative mechanism for the protection and preservation of environment.

It is to be accepted that SAARC ever since its inception has progressed on every issue despite hiccups, even then the rate of progress is very slow and in certain areas it has proved to be an utter failure. It is the politics between the nations that has dominated the stage of SAARC no matter how exuberant and progressive were the joint press statement release every time, the apathetical attitude of member nations had proved that it lacked conviction. The relations between the member nations were never cordial; it has been nurtured and harnessed by the mistrust among the member nations. India and Pakistan are at the loggerheads since their birth, India and Bangladesh are suspicious of each other's strategic interest, with India and Sri Lanka LTTE is an issue, Nepal has its own share of crisis with Maoist, these are some of the issues which SAARC has been unable to address and hence the core developmental issues have always taken a backseat. The stage of SAARC has become a place of launching propaganda against one another. The small nations like Bhutan, Nepal, and Maldives have concerns regarding their share of profit from the trade and are never eager to open their markets, besides they too have accused India for its hegemonic control over the association. A host of developmental issues including the environment issues have been relegated to reserves and the hopes are faint of SAARC becoming a group like successful counterparts European Union or ASEAN.

14

Environment Resource and Population Explosion in Indian Context

Prof. Rajeshwar Upadhyaya*

India is the second most populous country in the world. Every year an additional 17 million babies are born. The country's population according to 1981 census was 68.5 crore and it rose to 1 Arab by the year 2001. At present it is estimated to be in fast increasing order. Today India accounts for 15 per cent of the world population while its land area is only 2.4 per cent. The size of the earth is fixed and its resources are strictly finite, thus the difference between demand and supply of natural resources is damaging progressively. This has resulted in environmental degradation and shortage of easily available natural resources. Environmental degradation has now become a world problem and seminars and meetings to discuss the problem are being held on world level for sustainable development.

The Stockholm conference on Human environment 1972 and the international conference on population 1984 held in Mexico city reflected concern for sustainable development. The 1987 report of the world commission on environment and development, headed by Norwegian Prime Minister has used it as a key concept for developmental planning.

Although this is world wide problem but India has become a major victim of environmental degradation and pollution. The

* **Professor, Ex-Dean, Faculty of Education, Banaras Hindu Unitersity, Varanasi.**

problem has been drawing attention of Government, planners, educationists and even common people. Surveys and studies have concluded that for meeting out the basic needs of the people and improving their quality of life we should find out such approach which may not adversely affect the viability of the environment. The best approach to attain the goal of sustainable development may be the implications of interrelationships among pollution, environment, resources and development. Development is the ultimate goal that depends on the resources, may be natural and man made. Both the types of resources have their roots in the environment. Hence there should be proper understanding of the environmental problems and possible solutions to meet out the environmental degradation in this industrial age.

Major environmental problems may be enumerated as—deforestation, soil erosion, floods, ground water exploitation, threat to wild life, use of chemicals and pesticides, mining, water, ocean and air pollution, acidification, green house effect and ozone depletion. Here it seems pertinent to deal with the above mentioned factors cautiously so that we may proceed to improve the quality of environment and towards the issue of conservation and utilisation of natural resources.

Deforestation

Some poet has said regarding the importance of forest:

भारत में वन पावन तूही, तपस्वियों का तप आश्रम था,

जग तत्त्व की खोज में लीन जहां ऋषियों ने अभग्र किया श्रम था।

In ancient India forests were pious, free from contamination and safe place for meditation. Our Gurukuls (modern colleges and universities) were situated in forests, where teaching and researches were going on. In addition forests represent a well organised and highly evolved community of plants and animals. They are sources of materials of daily use such as food, timber, fire wood, wood pulp and fibre and medicinal plants. They are sources of many industrial raw materials and maintain ecological processes and life support systems.

The national forest policy of 1952 recommended that 33 per cent of the land area of the country should be under forests. Contrary to this, forests have been increasing assault since independence. At present not more than 10 to 12 per cent of the land is estimated to be under forest cover. The denundation of Himalayan and other hilly areas has led to soil erosion which affects river water quality and becomes cause of siltation of dams and reservoirs. Degradation of forests leads to the destruction of wild life habitats. About 135 species of plants need immediate care for their survival according to the Botanical survey of India. A study in Almora reveals that because of deforestation soil productivity has decreased and growing population in busy to maintain themselves through collection of woods and hunting forest animals. Thus we see that increasing deforestation has become the major cause of environmental pollution.

Soil Erosion

Soil degradation has serious economic and ecological implications for a country, specially for India where agricultural economy is considered to be the base. Soil erosion increases the silt load in rivers, dams and reservoirs. Kanwar 1980 has observed that siltation of dams reduces their effective life and ultimately causes serious economic and ecological implications.

Flooding

Loss of human lives and livestock and economic losses and damages to property are escalating because of floods. The pressure of population growth forces people to encroach upon flood plains. This disastrous input of floods magnifies as the size of population inhabiting flood plains and river banks increases. The average economic loss due to floods during 1971-78 was nearly Rs. 1000 crores annually. In Varanasi, where Ganga purification as launched by Late Prime Minister Rajeev Gandhi, many colonies have been established at the bank of River Ganga right from Assi to Bypass road. In such situation how can we imagine the eradication of environmental pollution? Such will be cases with other cities also.

Threat to Wild Life

More than 1200 species of birds and 500 species of mammals exist in the Indian region and include such animals as the elephant, deer, cheeta, lion and tiger. It is, however, pertinent to note that the population of wild life species are rapidly dwindling and many plants and animals are sources of many industrial raw materials and they balance nature in many ways.

Use of Chemical and Pesticides (Fertilizers)

India has tripled its cereal production by adapting modern means of agriculture. Chemical fertilizers are the most essential input for modern agriculture. At present the consumption of NPK (Nitrogen, Phosphorus, Potassium) in the country is 10.3 million tonnes. By 2000 the demand increased the 20 million tonnes to feed the growing population and will increase subsequently. Such a large input of chemical fertilizers will create diverse problems of environmental pollution which may be grouped into two categories:

1. Pollution caused during fertilizers production;
2. Pollution resulting from the use of fertilizers.

The present agriculture production is going to magnify the problems of environmental pollution in the coming decades.

Pesticides

The use of pesticides in our country has grown rapidly since independence. DDT (Dichloro Diphenyl Trichloroethane) BHC (Benzen Hexa Chloride) are the two main pesticides used in large quantities. The success of pesticides has been a mixed blessing because their use has serious environmental implications. Pesticide use, therefore, should be strictly controlled and regulated. The Government of India passed an Insecticide Act in 1968, but its implementation was very poor.

Mining

Mining and mineral processing have a great impact on environment. The ancillary and down-stream industries drawn by the mining activities, cause further deterioration of the environment.

Water Pollution

Water is the life-blood of the environment. Without water no living thing-plants, animal or man can survive. Our economy depends upon the availability of an adequate and usable supply of water. It is required by industries and the demand of water may increase to 16 times by 2025. The level of the ground water is going down day by day and at many places sufficient water is not available. As a result, inadequate hygiene, poor sanitation and discharge of industrial effluents have caused pollution of surface water sources. The rivers are polluted and extent of pollution is growing seriously. Some industries have established pollution abatement devices. We should be cautious about river pollution as far as possible. Ocean and coastal areas have also suffered from the pressure of growing human population. An oil cover on the sea surface blocks the free exchange of gases between sea water and the atmosphere and this causes, environmental pollution in many ways.

Air Pollution

Air pollution is caused by moving and point sources. Automobiles, ships, aircrafts, vehicles and rockets are moving sources, while industries and thermal power plants are examples of point sources of air pollution. Sulpher Dioxide (So2) is the most important among the gaseous pollutants. It is released by thermal power plants and other industries alongwith a large amount of flyash.

Ozone Depletion

It has recently been discovered that the protective ozone layer is getting eroded due to the impact of increasing human interference with the environment. The major cause for the depletion of the ozone layer is the world wide emission of man made compounds called chloro-fluoro-carbons (CFCs). The compounds are used in refrigerators-air conditioners, aerosols, sprays, for the cleaning of computer chips and in the making of rubber foam etc. CFCs ultimately find their way to the stratosphere where they chemically react with ozone molecules, breaking them down. Here in order to save human beings from the effect of uv radiation we should try to have control on excess release of CFCs.

Besides the mentioned causes of environmental pollutions there are other factors such as:

Acid Rain: which causes environmental acidification;

Green House Effect: Increasing carbon di-oxide will introduce sweeping change in the biosphere;

Noise Pollution: By vehicles, industries and machines;

Smoke Pollution: By brick kiln, transport etc.

Having described the kind of environmental pollutions we should have sincere efforts to improve the quality of environment. The primary steps in this campaign may be suggested as:

1. *Awareness of Environment:* Everybody should be alert and sensitive towards the importance of clean environment;
2. *Knowledge of Environment:* What environment means—How it changes—what problems it carries out of change;
3. *Attitude towards Environment:* Knowledge of social values, Development of positive attitude towards environmental balance, Motivation to eradicate environmental pollution;
4. *Participation in Environmental Improvement:* Group activity, Sense of Responsibility for preventing pollution and eradicating it;
5. *Skill of Environmental Improvement:* Skills should be developed for environmental protection and to solve the problems evolving from environmental degradation;
6. *Evaluation of Environment:* Time by time assessment of the effect of social, economic, political, aesthetic and Educational factors which cause environmental change;
7. *Hazards of Nuclear Power:* Transportation of radioactive-waves etc.

REFERENCES

1. Kanwar, J.S.—Save the Soil Save the Nation, *ICRISAT,* Hyderabad, A.P. India, 1980.
2. Khushoo, T.N. 'Environmental Priorities in India and Sustainable Development' 73rd Session, Jan, 1986, *India Science Congress Association, Calcutta.*
3. Ministry of Environment and Forest: *Annual Report 1988-89 Govt. of India,* New Delhi.
4. Sadik Nafis: Safeguarding the Future, New York, *UNFPA* 1989.
5. Varshney, C.K., Biosphere Reserves, *India International Quarterly,* 1982.
6. World Commission on Environment and Development, 1988—Our Common Future, *Oxford University Press, Oxford.*
7. Population Education—A National Source Book (Vol. 1), *NCERT, New Delhi,* 1991.

15

Environment

Different Issues

Dharmendra Kumar Mishra*

From the very beginning, the environment has remained the basic resource for the survival of human being and other living organisms. This makes it imperative that every one pays proper attention to the protection of environment. Therefore, men of every religion, race, caste and creed worship environment in many ways. Consequently, there were no problems of pollution of environment. In the primitive stage of the development of society, there was much sophistry about the metamorphosis of the primitive man into a civil order of life. Man was in a state of nature, a stage where he was one with nature. Man in a state of nature, wandering hither and thither without speech and industry, without home and hearth, was an equal stranger to war.

Man soon discovered innumerable modes and means of satisfying appetitive needs. Unfortunately, aided by rational faculty, man continued to attempt and adopt diverse modes of existence. Thus, man developed sophisticated ventures of life, a lust for power, property and dominance and his tendency towards self-preservation cut across its bounds into self-aggrandizement.

An appraisal of growth of culture and civilization has to be made from the perspective of the fullness of reality. In the fullness

* Senior Lecturer, Department of Law, M.B.S.P.G. College, Gangapur, Varanasi.

of reality, there is nature at large and the man enlarging his self at the expense of nature. In each moment of his life, he is imbibing one or the other of the resources of nature without ever returning any of the nature; bounties.

Thus, the gradual use of nature (natural resources) in an unplanned way by man started to disturb eco-system and ecology of the environment and this gave rise to major problems of the present era.

What is Environment

It is very difficult to define environment. Generally, it relates to surroundings. However, at present the environment has taken a more specific meaning and may be treated as covering the physical surroundings that are common to all of us, i.e. air, space, water, land, plant and wild life etc. There are different attitudes regarding environment.

The environment is defined as external conditions and influences affecting the life and development of organism. From this basic definition the human environment can be defined as a constant artificially created by man though yet dependent on nature. Depending upon the need, the external conditions and influences can include physical factors such as air, water, noise, soil and others.

Environmental Values

Environmental values are contained in individual or social beliefs about what is important in life and thus about the ends or objectives which should govern and shape public policies. The Royal Commission on Environmental Pollution in its 21st Report suggested about environmental values that certain of these factors may influence individual's attitudes to the environment. This may include such things as the environment as a vital resource for humans; the richness of biodiversity; and cultural, historical or social significance of the environment. Our priorities must be an integrated for environmental policy. Here, it is also important that opinion and attitudes must be distinguished as well as individual values of groups or sections of society as a whole.

Some values are closely concerned with environment. Environment is a store-house of natural resources. Man has been using these resources fro his benefit ever since his presence on this planet. In its enthusiasm to control and win over nature, the human race has used these resources indiscriminately.

Individual values can be sincerely held opinions that may fluctuate or be mutually inconsistent; at their worst, they are little more than consumer type preferences. These preferences gave to rise many issues regarding to the environment.

Different Environmental Issues

In the process of development, there are many issues involved with environmental pollution i.e., water, air, noise pollution and hazards waste etc.

Air Pollution

The Engineer's Joint Council defined air pollution as presence in the outdoor atmosphere of one or more contaminants such as dust, fumes, gas, mist, odour, smoke or vopour, in quanities or characteristics and duration such as to be injurious to human, animal life or to property or to interfere unreasonably with the enjoyment of life and property. The pollution of the local atmosphere from emissions has traditionally been easy to identify, such problems date back to the early use of coal in domestic fires which produced fumes and caused pulmonary infections and related lung diseases. With the advent of more complicated process, in the late eighteenth century, the problems of atmospheric pollution grew more severe. The main sources of air pollutions are industrial, transportation, agricultural, commercial and domestic heat and power, municipal activities and their fallout. The industrial revolution increased the use of coal to drive new machinery and more importantly, produced very high level of emission as a consequence of the 'alkali works'. These works are used the Leblanc process to produce soda, but the by-product of the process used meant the hydrochloric gas was emitted into the atmosphere which when mixed with water created a new phenomenon-acid rain.

Acid Rain

During the seventies, the problems of emission of smoke, dirt, dust and grit lessened, however, with the new approach to industrial processes the combination of sulpher-dioxide and other acids from power stations, industrial areas and traffic being combined with atmosphere produced not only rain but also acid depostits in the atmosphere made of ammonium sulpher particles.

Global Warming and Ozone-depletion

Global warming is one of the biggest environmental challenges now the world is facing. Since the past century, progressive industrialization has increased various gases in the atmosphere, which has as given rise to greenhouse effect. Consequently, the amount of ozone in the upper atmosphere which screens the earth from harmful UV-B radiation, has been deteriorated and there have been studies showing a 'hole' in Ozone lawyer. This also resulted in a climate change. Green-house-effect refers to the gradual warming of the earth's atmosphere due to the presence of green house gases which have warming influence on the world climate.

Water Pollution

Water is one of the primary essentials of society which is used for cooking, drinking, washing, cooling, irrigation, industrial processes, waste material disposal and many other purposes. Water quality is every body's concern and almost all countries have made it a national goal to eliminate discharge of pollutants into natural water bodies.

Water pollution increases from many other sources including commercial, domestic, industrial, waste water, agricultural practices and shipping. But the major pollutant is sewage.

Noise Pollution

Noise is an unwanted sound. Sound is generally, measured in terms of intensities or pressure levels, which is nearly expressed in desibles (dB). Excessive sound levels of any type will lead to hearing losses. Noise pollution level and its harmful effect to human being is as follows:

30 to 60 dB	—	Stress, Tension and Illness.
60 to 90 dB	—	Disturbance in stomach and high B. P.
90 to 120 dB	—	Grevious hurt to health and ear
120 to 150 dB	—	Painful effects irreparable loss of hearing

The major sources of noise pollution are transport vehicles including railways and aeroplane, electrical gadgets, machines and mills etc.

Pollution by Hazardous Wastes

The problems of waste generation is one of the by-products of industrial development. Wastes are defined as substances, solutions, mixtures or articles for which no direct use is envisaged but which are transported for reprocession, dumping elimination by any method of disposal.

Hazardous wastes may be explosive oxidizing, toxic, carcinogenic and bio accumulative and may cause respiratory problems, cancer and other serious diseases to the human kind. It is quite difficult to dispose off safely most of the toxic industrial by-products. It is *sine-qua non* of modern industrial development that environmental pollution in some measure and form is indispensable. In the initial stage of industrialization more importance was attached to stepping-up of production and industries—regardless of the need to install pollution-free technologies. This has happened all over the world and is being continued.

Precautionary Measures to Protect Environment

All the environmental problems can be solved through technology, public awareness, rules and regulations. The Govt. of India and several States have made administrative and legislative arrangements in order to ensure environmental protection. Near about two hundred central and state laws are in existence to protect environment. There are certain major central laws i.e. the Wild Life Protection Act 1972, the Water (Prevention and Control of Pollution) Act 1974, the Air (Prevention and Control of Pollution) Act 1981 and the amended Environment (Protection) Act 1986.

The team approach awareness amongst people may give total solution regarding environmental pollution by using advance technology. The role of the articulate middle class assumes importance in the protection-process. Generally, one who possesses the scientific or technical knowledge and the skills, can advocate effectively such problems. In the environmental sphere itself pollution monitoring and assessment bodies necessarily require scientific or technical knowledge and application of much techniques. Whereas participatory research may be adequate on issues like access to forest resources etc. There are many instances of the activists losing their credibility became of uniformed statement on technical aspect. A new class of people called "people science" has emerged very much as part of the environmental movement which comprises trained personal capable of analyzing technical aspects of a problem from a people centered perspective.

Thus, the general public is becoming aware of the need to protect themselves and environment even though, laws have been passed for the protection of environmental pollution, the enforcement of the same has been tardy, to say the least. In this connection, some suggestions may be given as under:

1. The environment protection machinery both at the center and at the states level should be strengthened and the machinery should be provided more teeth for strict implementation of environmental protection legislation;
2. There should be personal accountability for their lapses, negligence, resulting in damage to the environment by all pollutants;
3. The principle of sustainable development should be strictly adhered to by the policy making bodies;
4. The principle of 'polluter pays' should be given the statutory status;
5. Eco-friendly technology should be developed in designing the machines to avoid pollution;
6. Change is needed in life styles of developed and rich countries vis-à-vis consumption patterns of all countries.

Environmental pollution control is the system of processes which limit the impairment of the quality of water human's use, the air they breathe and the land that sustains them. It includes the means to control the physical energies of ionizing radiation, non-ionizing radiation, sound, air-pressure changes heat and cold. It aims at fulfilling three objectives. The first is to protect people from physiological damage from pathogenic organism, from toxic chemicals and from excess use of physical energies. The second is to spare human annoyance, irritation and discomfort from offensive conditions in air, in water and on the land. The third objective is to safeguard that balances in the Earth's ecosystem to conserve natural resources and it should be the primary goal of environmental pollution control.

REFERENCES

1. Ambasht, R. S. and Ambasht, P.K., *Environment and Pollution,* (An Ecological Approach), Student's Friends & Co. (1990)
2. Bartelmus, Peter, *Environment and Development,* Allen Unwin Inc. London (1986).
3. Ball's and Bell's, *Environmental Law,* Universal Law Pub. Co. Pvt. Ltd. (1991).
4. Dix. H.M., *Environmental Pollution,* John Wiley (1981)
5. Gore, Al, *Earth in the Balance: Ecology and Human Spirit* (1992).
6. Rao, C. S.: *Environmental Pollution Control. Engineering,* Wiley Eastern (1991).
7. Trivedi, P. R.: *Encyclopaedia of World Environment,* Vol-2, A.P.H. Pub. Corp. New Delhi, (1999).

Legal and Constitutional Framework

16

Indian Environmental Law

The Finished and Unfinished Agenda

Prof. C.M. Jariwala*

I

INTRODUCTION

India has travelled a long way from *Dharma* of environment to the Environment (Protection) Act, 1986 and further thereafter. The Indian environmental law, during the above journey saw at time smooth sailing but had also road blocks and cross roads. Its changing shape brought in new developments, requiring new tools and machineries to prevent and control the environmental pollution. In this journey, however, certain agenda got cleared and other remained unfinished and uncared for. On the otherside of the coin is the sad story of environment, moving from natural to deadly polluted. Today the environment is in the danger of further degradation, the living species are on the verge of extinction and what not. The unplanned, uncontrolled and one side development process has become a way of life and a first class enemy of the environment. The world community has now realised the need of a sustainable development where a balance could be maintained between the preservation and promotion of good environment and the development plans. But in the developing and under-developed

* Retd. Ex-Dean & Head , Faculty of Law, Banaras Hindu University. Varanasi.

countries, it has become difficult to please the two ends: environment and development. India is also facing this dilemma. The selfish and vested interests only care for more unplanned developments, getting more profit and earning more riches at the cost of environment. The government, to please this constituency, has also learned in favour of their side. It is on the initiative of the environmental activists and protectionists that the judiciary did not approve of this move. Thus one finds a continuous battle between the *bhakshaks* and *rakshaks* of the environment.

On the environmental scene, it has mutli-facets and many components. Each one is having its important position in the protection, preservation and improvement of the environment. Every aspect of environment needs legal control for its survival and further improvement for the generation to come. In all these positive, negative or inactive responses the man is responsible, in his individual, corporate and collective strength. To prevent and control their anti-environmental activities, it is said that the Indian Legislature has enacted a large number of laws, numbering more than 200. The bright and dark sides of the entire journey can be assessed by examining their contributions to protect and improve the environment and also their failures in their goal of environmental excellence. The present study takes an overview of the currents and cross currents in the changing dimensions of the Indian environmental law and points out its demerits and necessary reforms therein and the unfinished task before the Indian environmental law.

II

ANCIENT INDIAN HERITAGE AND CULTURE

It has been said that there are evidences in the ancient literature "to prove that the society in the ancient time paid more attention to the protection of environment than what we think today"[1]. Every constituent of environment is given an important place in the literature of different religions which flourished in ancient India. The rivers were considered as embodiment of Goddesses and their water was considered very pious. It was not only having self-purifying capacity but also drinking their water was supposed to lead one to heaven. *The Atharva Veda* describes the quality of water (*Apah*)[2]. *Charaka Samhita* enumerates the characteristics of polluted water and narrates its ill effects on human beings[3]. In Kautilya's *Arthasastra* one finds graded punishments prescribed for the pollution of water[4]. Similar treatment was given to air (*Vayu*) and other components of environment in these scripts. The Mahabharata has adopted a different way of explaining the importance of different contents of environment. Each component is identified with different organs of God. For example, the earth is His flesh, sea is the blood, sky is His abdomen, air is His breath, agni is teja, rivers are his nerves and the sun and moon are his eyes[5].

A prominent place was given to plants and animals. Dwivedi gives a list of plants and animals whose names were associated with God or Goddesses. For example, Lion-Durga, Elephant-Ganesh, Serpent-Shiva, Peacock-Sarswati monkey- Rama, etc. *Vata*-Brahma, *Tulasi*-Lakshmi, *Ashoka*-Buddha, *Neem*-Sitala, *Pipal*-Vishnu, *Kadamba*-Krishna, etc.[6] Thus the animals and plants were accorded reverence and destroying the trees or killing the animals were considered a sin and it was subject to various punishments. The influence of Islam, Christianity, Buddhism or Jainism in India did not change the basic notions though their environmental philosophies might have been interpreted differently. Dale Jamieson's work the *Culture, Religion and Environmental Ethics* and R.P. Mishra's edited *World Environment and Environment,* Hindu Religion also support the above direction.

It is interesting to note that the ancient Indian law takes cognizance of not only the prevention and control of pollution but also cave for the proper use and reuse of natures gifts for the betterment of environment. This is what *Isavasyopanishad* teaches. A further teaching in the literature is 'live in harmony with nature', a concept which is reflected in the modern India in the Directive Principles of State Policy and the Fundamental Duties of every citizens provided under the Constitution of India, Articles 48A and 51A (g) respectively.

An indepth study of the literature would further reveal that the ancient Indian law had many commandments on dos' and don'ts which brought in the idea of mandatory duties. In this atmosphere, the right conscious approach could hardly flourished. At time, they provided a piece of advice and informations as to how best to achieve well being of the humanity. One finds also consequences of pollution elaborately explained and remedies to cure pollution off-shoots and ways to enhance the quality of environment were also clearly enumerated.[7] The present discussion brings home a journey from commands on preventive measures to internalizing protection and improvement of environment in the common life style.

The above literature, thus, discards the theory of the academics who take the stand that history of the Indian Environmental Law begins with the Stockholm Conference, 1992. On the contrary it supports the thesis that in ancient India a well defined eco-dharma existed which was given obedience by the people. There were two reasons for following the commandments: one, the fear of punishment, and two, the environmental culture, traditions and values had been internalised by the people of that age in their conduct and behaviour, becoming a part of their life style. But unfortunately today these elements have been side tracked in their priorities and the result is that the Indian environment is subject to continuous assaults and rape.

III

SPECIFIC LAWS OF YESTER MILLENNIUM

The Tiwari Commission in its Report has mentioned 200 legislations dealing with environment. But if one takes the stock of legislations dealing with all components and constituents of environment, the number would multiply many times. Our concern here is to concentrate on the law modalities for the eco-friendly use and misuse of the environments and its components, the culprits and victims of environmental pollution and a safe course for the future generation. A bird's eye view of the history of the yester century's environmental legislations show that except in few cases, the legislations cared for developmental activities more. The Stage Curriage Act, 1861 down to the Inflammable Substances Act, 1942 at the Central/Federal level and the different exercises at the provincial and grassroots levels support the above conclusion. More productions, more developments and more riches were the goals of the laws relating to industries and industrial development. The reasons for the above sea change in the scenario were mainly two: one, the exploitative greed of the Britishers over the Indian natural resources; and two, decrease in the ancient India heritage and culture and growing hold of alien culture.

The above description should not be taken to mean that the concern for the legal protection of environment had no place in the British *Raj*. In this jungle of eco-unfriendly legislations, the forests and wild life and animals laws, found a place on the statute books but the apathy was that they could hardly became functional to march forward towards the desired goals. The legislation which deserves attention is the Indian Code, 1860 which made provisions for the pollution of water and atmosphere under Sections 277 and 278 respectively. The former section makes it a punishable offence if one voluntarily corrupts or fouls the water of the public spring or reservoir and makes such water less fir for the purpose it is ordinarily used. The ordinary use of that age included: domestic, agricultural and industrial purposes. Section 278 talks of vitiation of atmosphere of any public place affecting the public health. These provisions had a very narrow vision and hardly paid any attention to its involuntary use and consequences.

The aforesaid offences could have a very wide repercussion, including death but the liberal criminal sanction was only three months imprisonment of either description and/or fine to the extent of Rs. 500/-. The same criticisms apply to the offence of atmospherical pollution with the difference that in this case section 278 takes care of its consequence but only on human health and the penalty is too liberal, fine to the extent of Rs. 500/-. The unfortunate part of these criminal law provisions was that they were hardly pressed in for the purpose for which they were enacted. This gets support from the case law under the above sections.[8] The victims of the aforesaid acts were generally the poor who had no courage to strike against the riches. Thus the few silver lingings in the latter period of history were hardly allowed to dawn on the horizon of the then Indian environment.

IV

ENVIRONMENTAL LAW IN PARTICULAR

A. The Constitutional Contribution

The unconcerned approach towards the environment and care for development in the British *Raj* did not change with the changed status of India, 'that is Bharat'. The Free India's constitution, the Constitution of India, no where talked about the environment: its protection, preservation and improvement. It conceived a right oriented society, a developmental approach where people were busy in claiming and enjoying their fundamental rights to property, business, trade, etc.

The Legislature took care not to violate the fundamental rights and the judiciary also, in its dynamism, tried to give protection to the aggrieved persons and enforced their rights. The result was that the development process excellerated but the environment and duty towards environment hardly raised their head in this scenario. Thus there was, on the one hand, deterioration in the existing state of Indian environment; and on the other, fading of environmental culture. The committed international environmental agreements also remained uncommitted and unenforced.

The mistake was realised after the Constitution of India had celebrated its Silver Jubilee. The change of India to the 'Socialist' status in the year 1976 saw the dawn of environment vision in the Constitution.. It put back in the constitutional document the lost glory of duty conscious citizens and the State as well. Article 48A was added to the Indian Constitution by the Constitution (Forty-Second) Amendment Act, 1976. A new provision was introduced under the Directive Principles of State Policy adding one more fundamental principle in the governance of the country, a matter of State's duty. This article imposes an obligation on the State that it shall endeavour to protect and improve the environment and to safeguard the forests and wild life of the country'. However, article 37 makes this provision not to be enforced by any court, which means that such provision is a show piece in the Constitution, But, the Rajasthan High Court has put a new shade in this so called dormant provision by evolving a fundamental right to live in clean environment.[9] Since 1976 Parliament has from time to time enacted laws or brought in reforms in the existing laws to protect the water, air, wild life and animals and forests.

The right oriented perspective held firm ground in pre and post-Independence period but it was realised, that it created many road blocks and cross roads for the environment of the country. The imbalanced development needed revival of the duty concept. Article 51A(g), a part of the newly added Chapter of 'Fundamental Duties', embodies the fundamental duty of every citizen of India 'to protect and improve the natural environment including forests, lakes, rivers and wild life, and to have compassion for living creature'. When this fundamental duty was imposed in the year 1976, the literates and illiterates were illiterate of environment. And, therefore, imposing abruptly such duty did not bring the desired results and the citizens continued to harp on their fundamental rights to exploit the natural resources and the environment in general. However to curb such activities, a fundamental right to a livable environment was given birth in article 21 by the Indian judiciary.

It may be pointed out that in case of both the aforesaid articles members of Parliament at the enactment stage advocated to provide enforcement teeth with rigidity to this provision. But their suggestions were not accepted. And, therefore, when one takes the stock of their output, it is not encouraging. However, one thing may be said that there is no constitution of the world which provides for a *trimurti* (three dimentional) of environment: obligation, duty and right, the pride of our nation. Inspite of such advances, there are still undone agenda which includes: Firstly, there is no clear source of legislative power relating to environment. The Constitution (Seventy-fourth Amendment Act, 1992, after more than four decades of India' Independence, gave Gandhi's dream of *Swaraj* at the grassroot level an operational force through the newly added provisions under Parts IX and IX-A read with Eleventh and Twelfth Schedules to the Constitution of India. The Twelfth Schedule to the Constitution in Entry 8 specifically talks of power of the municipalities relating to 'protection of environment and promotion of ecological aspects'. But it has to be seen whether the different States in India would allow the local self-government to walk on their own. Secondly, the provisions for 'Tribunals for other matters', in article 323B needs a place for tribunal for environmental disputes as well. This will give sanction of the Supreme Law of the land to the environmental dispute settlement mechanism. Thirdly, the fundamental right to environment must find a place in Part III of the Constitution of India so as to maintain a balanced approach in enforcing those fundamental rights[10] and the constitutional rights[11] whose enjoyment resulted in unfriendly-eco output. Fourthly, separate financial avenues have to be brought in the financial relations between the three tier system: Union, States and local levels. This will, to a certain extent, ease out the so called State bankruptcy in handling environmental matters and the polluters would not go scott free and continue to pollute the environment with no accountability and responsibility. And last but not the least, the Constitution of India must provide special provisions to deal with eco-disaster which requires emergency attention at the national level. The Bhopal mass disaster reminds us for such a need.

B. The Statutory Exercises

The legislative care for the protection of environment from pollution has a rich history in India The specific attempts start with 1974 and continues right upto the year 1997.[12] Parliament started with the Water and then Air and finally provided legal control on the protection and improvement of environment. The Central Government, in the exercise of the rule making powers conferred by the relevant Acts, made rules to provide detailed provisions on the matters which were left to the rule making authorities.[13] The Bhopal mass disaster brought new problems to be solved by the law provisions to provide immediate relief to victims of such catastrophe and also a forum to compensate the losses suffered thereby. For this Parliament enacted the Public Liability Insurance and the Environment Tribunal legislations.

An attempt is made in the following pages to provide a bird's eye view of the legislative exercises mentioned above. The discussion is divided under the following selected heads:

1. The Vision

The three main environment legislations: the Water Act, Air Act and Environment Act, were brought in for prevention, control, abatement protection and improvement of water, air and environmental respectively.

The Act of 1974 was passed to give effect to the wishes of some of the State Legislatures who opted for a Central legislation on the matter of common concern of two or more than two States. The water has intrastate and interstate dimension and if it is polluted it involves not just one State which pollutes the water but the other States as well who receive the polluted water. Such problem needs a central legal control. But the problem of this technique is that the States, who are more developed than the others, may not adopt such enactment and at the same time may and may not have their own Water Acts.[14] Thus at the initial state of the prevention and control of water pollution this was left at the sweet will of the concerned State Legislature.

On the other hand the Air Act, the Environment Act and also the Tribunal Act were passed to implement the decisions taken at the Stockholm Conference, 1972 for the preservation of the quality of air and protection and improvement of environment and prevention of hazards to human beings and other living creatures, plants and property. In case of the last legislation the purpose was to provide a mechanism to determine and award compensation for damages done to persons, property and the environment while handling hazardous substance.

Efforts were made by the Central and State Governments to see that the above legislative exercise come out of the Statute books and could touch atleast the boundary of the implementation regions but the progress is not too responsive. In case of the Tribunal Act, 1995 the sad story is that though a period of long six years have passed, the Central Government has yet to issue any notification for its commencement.

What is the result ? The environmental legislations it may be submitted, have very high sounding objectives but their operations, since their commencement, have hardly moved forward in the directions of the goals set by Parliament, a retrograde step to the democratic aspirations to protect and improve the Indian environment.

2. The Mechanisms

(i) Authorities

The Acts of 1974 and 1981 provided for the Water and Air Boards to prevent and control the pollution of water and air at the Central and State levels. Their functions are now performed by the common Central and State Pollution Boards. Under the Environment Act, 1986, the Central Government is given power to constitute authority or authorities. The Department of Environment, Forest and Wildlife is created to fulfil the State's constitutional obligation under article 48A.

Then there is the Environment Appellate Authority created under the national Environment Appellate Authority Act, 1997 which may hear appeal against any order not granting or conditional granting environmental clearance for location or processes of any industry. But, the constitution of the Authority is not spelt out in detail rather it is left to its own will of the Central Government. The Act should have provided the sources of membership instead of the government distributing political rewards and favours as it may deem fit.

These authorities have to perform very important functions which include: fixing of environmental standards, giving consent to operate industrial process, environment impact assessment, nation and state-wide environmental programmes to disseminate informations etc. Each function in itself requires lot of money and manpower which the concerned government always says it is difficult to get. And, this is the reason that the above functions have not functioned with that urgency as the protection and preservation of environment needed. Here comes the requirement of a monitoring agency, an ombudsman, to keep the authority on its toe and making it accountable for any lapse or undue delay in action. The Second aspect which needs reform is the constitution of the authority which is full of government's representatives, nominees or officials or in some cases left to the sweet will of the government. It is necessary that people's participation has to be given top priority in environmental decision making. The laws providing the rule making power must ensure time schedule for the performance of each programme. Last but not the least, the clearance of environmental matter has to pass through different channels. This takes it own time and different application of minds giving scope for more different decisions and avenue for corruption. There are occasions when too many cooks spoiled the food. The need of the present software age is: one window clearance.

(ii) Dos' and Don'ts

The environmental laws prescribe certain conditions and prerequisites to be and not to be followed to prevent and to protect and improve the environment. The positive aspects include: to follow the prescribed standard, to disseminate required information, to follow the directions issued and abide by orders issued by the appropriate authority. The legislations commands not to do: not to discharge polluted water in wells and places of storage of water, no new outlet or new discharge avenues be opened without permission, desist from causing pollution, not to obstruct the authority/officer from performing its function, not to hide information on crucial issues related to environment. These dos' and don'ts are the basis of offence under the laws in case of its mis-performance or non-performance.

(iii) Penalties

The environment laws have prescribed penalties for the performance of prohibited or restricted acts or not fulfilling certain required conditions. First type of actions include power to close down, prohibit or regulate industry or industrial processes or stoppage or regulation of the supply of electricity or water supply or any other service to the estiblishment, prohibition or restriction on handling of hazardous substances and location of industry, require remedial measures to be adopted in case of accident causing environmental pollution.

Coming to penal sanction. Parliament has moved from liberal to rigid sanction. Under the Water Act, 1974 the penalty for non-compliance of any direction was three months imprisonment or fine which may extend to ten thousand rupees or with both. However, in case the failure continued then the penalty could extend to five thousand rupees for every day of its continuance. In case certain orders were not complied with then the punishment could extend from of one year and six months to the maximum of six years imprisonment and fine. The penalty for continuance of the offence was same as provided above. The Environment Act 1986, on the other hand, incorporate heavy penalty to the extend of six years imprisonment or fine which may extend to Rs. one lakh. In case of its continuance the punishment may extend to seven years imprisonment.

It may be pointed out that these penalties have been diluted by the Acts of 1974, 1981 and 1986 in favour of company and government departments if the offence was committed without knowledge or the eco-cider adopted due diligence to prevent the commission of the offence. This is nothing but leaving the big polluters and catching the small ones. Such is the scene at the international level also where the over-developed nations sit in the favoured nation block. Can this approach save the environment from further degradation ? In order to safeguard the environment, it is suggested, that the legislature cannot and should not discriminate between haves' and haven'ts. Everybody holds the environment as a trustee for the future generation. And, therefore, they should be equally accountable for their adverse actions irrespective of their position and status.

It may be pointed out that there were cries in Parliament to handle the eco-rapers with severe penal sanction. But Parliament was contended with the above penalties so that the Indian environment was not further degraded. But the tragedy of the liberal to rigid penal sanction has been that it could hardly see the dawn of its infliction and, as such the deterrent punishment remains in the cold storage of the Stature book, frustrating the will of the people. It may be mentioned that penal sanction is not the only solution to control and prevent the evils of pollution. If Mr. Anderson of the Union Carbide is subjected to the criminal sanction would that balance the losses suffered due to the Bhopal mass disaster ? The answer shall be only no and nothing else. The anti-environmental activities require for their control, not only mechanisms preventive and prohibitory in nature but also incentives, recognition, black-listing the products and the polluter, remedial actions and rehabitutory plans. This second type of actions have presently no place in the aforesaid environmental laws. It is the time that the Indian laws must provide such actions as well to duty protect and improve the environment[15].

V

JUDICIAL ENVIRONMENT LEGISLATION

There are criticisms and counter-criticisms against the Indian judiciary for their role in the administration of the environmental justice: it is 'not competent to handle environmental litigation's having' great technological complexity[16], or its 'highly innovative, 'super innovative' and most powerful court in the world'.[17] During the last five decades, the environmental justicing and justice have seen many ups and downs[18]. The courts in their expansive justice role made important contributions in the field of environmental law which include: evolving the fundamental rights to live in clean environment, environmental education and information through article 21 of the Constitution of India.[19] These fundamental rights will require the courts to maintain a balance between the aforesaid rights and the fundamental right to development (Arts, 19(1)(g), 300A, 301, etc.) These fundamental rights will allow the petitioners to make responsive the irresponsive and misresponsive authorities. It is pity that they have to be told time to time and again about their constitution obligation to 'protect and improve the environment'.

But the major hurdles and questions are: the environmental illiteracy in the poor masses; how many can afford to move the high court and/or the Supreme Court; how many can withstand the wrath of the industrialists and the other big fishes in the pollution game and politics over and above this is the corruption in the environment jurisprudence. Then there is the problem of time consuming processes and also of enforcement of judicial verdict when the tool of contempt of the court has not made much headway.

If one takes the stock of cases where the courts propounded or enforced the fundamental right to live in a clean environment and the other aforesaid right, the result will be depressing. The Doon Valley[20] is one of many examples where the judicial justice has yet to reach to its desired direction. Today the Valley is very much denuded of its greenery by the eco-rapists. And, therefore, there is a need to have a statutory monitoring authority to see that the judicial output see at least the light of dawn; and second, the court must fix accountability on those who flout the court's orders/directives. The habitual disobedients must be handled with iron hands. This is one of the ways to repose the faith of We, the People of India' in the judiciary.[21] It is further necessary that the environmental literacy and legal aid must find an effective place in the environmental jurisprudence. The justices have their own philosophy and at time did not hasitate in favouring pro-developmental approach, resulting in far reaching adverse consequences on the environment. The *Banwasi Seva Ashram* case[22] is one example where such approach has affected greenery of the Vindhya Valley. On the other hand, the judiciary has also important innovations by evolving eco-friendly principles. These principles have not only brought in new vision to environmental norms and laws but also opened doors for more fresh winds in the protection and improvement of the existing degraded environment. These principles, for example, include; 'polluter pays' which means that the cost has to be beard by the person who pollutes the environment and there is no scope for any exception—unintentional or accidental act; 'precautionary principle', where it is necessary to take precautionary measures to anticipate, prevent or minimise the cause of pollution and mitigate its adverse effect. This principle does not allow any delay in arresting pollution and further that it is the polluter who has to prove that his action was eco-friendly, instead of the victim to show that the action caused pollution,[23] intergenerational equity which supposes that the natural resources are no man's property and they are held by the present generation as a trustee for the coming generation so that the incoming generation do not get degraded environment.[24] Here the court is evolving the right of an unborn. In all these principles the question is: how for these principles will go down to the behavioural regime of the common men? Will the ecociders follow these commandments? They, therefore, need the umbrella of legislation.

The courts have also evolved new set of provisions and procedures which the existing law have yet to provide.[25] It is time that the legislature must take note of the flaws and gaps in the law pointed by the judiciary and, the existing laws have to be tailored or made to accommodate the changing needs. The judiciary in order to deliver judicious justice has evolved easy justice processes and made available more remedies. The Public Interest Litigation allows any public spirited person or organisation to move the court instead of the long standing rule of 'affected person only'. In the matter of environmental pollution there be many victims some of whom may not know its ill effects, or may not be aware of their rights and may not even afford to move the high courts or the Supreme Court. In such a case the non-governmental organisations and public spirited persons have knocked the door of the court to get environmental justice.[26] But at time this process has been unsuccessfully initiated by the vested interests to serve their own purpose.[27] The judiciary must handled firmly those who come to the court with ulterior motive. Coming to remedies, the courts have provided compensation[28], and imposed fine,[29] allowed criminal proceedings against the polluter[30]. The reformatory and rehabiliatory processes were also ordered to prevent further pollution and set back the house in order.[31]

The courts exposed the anti-environmental activities of the industries[32], the politicians abetting pollution[33], executive illiteracy and inaction.[34] But the problem is that these culprits are hardly made accountable and this has encouraged their eco-unfriendly exercises. The Supreme Court of India has the inherent power under which it can see that irresponsive persons or authorities are taught good lesson. In this sensitive region the court cannot and should not keep its sword blunt.

The environmental pollution, its protection and improvement are complex matters and at time involves science-techno-niceties. Though the judiciary tried to handle these complex issues,[35] yet it has shown stresses and strains in weighting the balance.[36] There are occasions when the court was flooded with reports of the committees and experts viewpoints [37] to understand the complexities and decide the issue. In the every day innovations in this field and the enormous literature coming up every moment, it is necessary that the justices have to be regularly exposed with these developments and the high courts and the Supreme Court must have an academic and research cell to help the court in this regard.

VI

WHERE DO 'WE' GO NOW?

The history of environmental pollution and litigations, not only in India but also throughout the world, reveals that most of the

Victims were the helpless and hapless poor men. The Hiroshima, Chernovly and Bhopal mass disasters are evidence of this fact. The rich eco-cider silently raped the environment and left its aftermath in a miserable condition, deteriorating environment and increasing sufferings to the living and non-living beings. How long this *Mahabharat* can must be allowed ? The clear answer is, no longer; otherwise, it will destroy everything, *sumpurna vinash* as predicted in Bhagawat Geeta. The eco-enemies must realise that they are digging graves not only for themselves but also for their future generation. It is time that the Government of India must constitute an environmental council, giving due representation to both the sides along with members who, care for sustainable development. It shall suggest from time to time proper ways and means to safeguard the environment from further degradation and also to recreate its lost glory. Parliament shall then shape the existing or a new mechanism in this regard, instead of allowing politics with in environment. This should be the priority agenda of all the parties.

Parliament in the exercise of its constituent power must provide an integrated approach in the Constitution of India to protect and improve the Indian environment. The right and duty of 'We, the People of India' towards the environment and the obligation of the State in this regard must find a separate place in the 'Fundamental Rights' chapter in such a way that the aforesaid *'Trimurti'* along with other fundamental rights, duties and obligations build up an amicable rather than inimical relationship, leading to ecofriendly atmosphere. Further the vague situation in the source of legislative powers be done away with by putting 'Environment' in the concurrent judisdiction of both Parliament and the State Legislature.

The environmental legislations have well meaning objectives but these objectives must see the light of the day instead of lying in the store house of the Statute books. The legislations must provide accountability of those on whose shoulder is the responsibility of making the objective part operative. An authority like Ombudsman be constituted to report Parliament from time to time about the progress made in this regard and Parliament must take action thereon. Each pollution requires different handling by legislation and specific rules, however, it must have an uniform approach for an integrate treatment. The legislations cannot simply depend on criminal sanction but they must provide eco-friendly measures so that the *bhakshaks* are properly dealt with and *rakshaks* are encouraged. Further an environmental research cell be constituted to collect necessary date and informations on the operative part and the part which still remains inoperative so that the existing laws may be reformed accordingly and new legislations can be made.

The judiciary has done commendable job in making environmental judicial legislations. But the pity is that its fruits have not reached to the relevant components of the environment. Though environmental justice might have been done but it remains to be seen and felt. It is time that the judiciary must equip itself with such mechanisms that its decisions/orders attain the desired objectives. The fundamental rights to legal aid and speedy justice must dawn on the environmental litigations. The judiciary is under the constitutional obligation to make all 'endeavour to protect and improve the environment', as such, individual bias, prejudice or philosophy should not be allowed to play any role in environmental justicing.

So what we need? The Indian Environmental Law cannot and should not remain static but it must change with the changing dimensions of environmental degradation. The second and one of the most important component is the growth of environmental literacy with a touch of ancient India culture which must reflect in the conduct and behaviour of every Indian. This will open all out efforts and commitment from every corner to save the Indian environment from further degradation and a way to revive the lost glory of the Bharatiya environment.

REFERENCES

1. C.M. Jariwala, Changing Dimensions of Indian Environmental Law in *Law and Enviornment* Edi. P. Leelakrishnan, 2 (1992) See an interesting Work, Dwidedi, O.P. and Tiwari, B.N. *Environmental Crisis and Hindu Religion* (1987). Also See Dwivedi, O.P. Classical India in *A Companion to Environmental Philosophy* Edi. Date Janieson (2000); Environmental Ethics A Dialogue of Cultures, Edi R.P. Mishra (1995).
2. *Atharva Veda,* bBook IX, 2, 1.
3. *Charaka Samhita, Vimanastanam* 111, 6.1. See also *Mahabharat Rajdharmanuasasanpurva,* Chapter 16.6.1.

4. *Kautilya's Arthasastra*, Book 11, 36.145. *Ghora Narak* (hell-fire is prescribed in *yajne Valkyesmrti, Achryadhyaya*, v, 180). See 14 commandments in *Prayaascittatatva*, 1.535.
5. *Mahabharata Moksa*, 182, 14-19.
6. Dwivedi, O.P. and Tiwari, B.N., *Enviornmental Protection in World Religion and Environment.*
7. See, for example, *Visnusmriti*, 23.38-46, *Charak Samhita, Vimansthan*, 3.2.
8. C.M.Jariwala *supra* note 1.
9. *L.K. Koolwal v. State of Rajasthan*, A.I.R. 1988 Raj. 2
10. Articles 14, 19(1)(d),(e),25-28,29etc.
11. Articles 300A,301, etc.
12. The Water (Prevention and Control of Pollution) Act, 1974, the Water (Prevention and Control of Pollution) Act, 1981; The Public Liability Insurance Act, 1981; the Environment (Protection) Act, 1986; the National Environmental Tribunal Act, 1995 and the National Environmental Appellate Authority Act, 1997.
13. *See* for Example: The Water (Prevention and Control of Pollution) Rules, 1975; the Air (Prevention and Control of Pollution) Rules 1982, 1983; the Environment (Protection Rules, 1986, etc.
14. Art. 252 talks of consent of State which will imply not joining the Central control or withdrawing consent at a later stage, opting out from the operation of Parliament's enactment. The State of Maharashtra had adopted Aya Ram and Gaya Ram approach in case of the Water Act, 1974.
15. *See* Art. 48A.
16. *See*, U. Baxi and T. Paul, *Mass Disasters and Multinational Liability; The Bhopal Case*, 178, 183 (1986).
17. *See*, the Affidavit submitted by Dadachangi in U. Baxi, *Inconvenient Forum and Convenient Castastrophe; The Bhopal Case*, 76 (1986); the Affidavit submitted by N.B.Palkhivala, in Baxi and Paul, *Mass Disasters and Multinational Liability; the Bhopal case*. 224, 226 (1986).
18. *See*, for the Fifty Years Account in C.M. Jariwala, the Directions of Environmental Justice: An Overview in *Fifty Years of the Supreme Court of India—Its Grasp and Reach*, Edis Verma and Kusum, 469-494 (2000).
19. *Rural Litigation and Entitlement Kendra Vs. State of U.P.*, A.I.R. 1989 S. C. 594; *L.I.C. of India Vs. Manubhai Shah*, A.I.R. 1993 S.C. 171; *M.C. Mehta Vs. Union of India*, A.I.R. 1992.S.C. 383.
20. See for example, *Rural Litigation and Entitlement Kendra Vs. State of U.P.*, A.I.R. 1989 S.C. 171 where the Supreme Court of India enforced the right of the Kendra against the illegal mining activities in the Doon Valley.

21. A statistical compilation of data of the responses of the common men can be an interesting study to find out the increasing or decreasing credibility of the judiciary in India in the eyes of the masses. But here it is submitted one cannot have a very high positive hope of a pleasant outcome.

22. *Banwasi Seva Ashram Vs. State of U.P.,* A.I.R. 1992 S.C. 920 where Justice Bhagwati titled the balance on the development side.

23. *Vellore Citizens Welfare Forum Vs. Union of India,* A.I.R. 1996 S.C. 2715. *M.C. Mehta Vs. Union of India,* A.I.R. 1997 S.C. 734.

24. *A.P. Pollution Board Vs. M.V. Nayudu,* A.I.R. 1997 S.C. 812, 824. *Indian Council for Enviro-legal Action Vs. Union of India,* A.I.R. 1996 S.C. 1446.

25. See, for example, *Centre for Social Justice Vs. Union of India,* A.I.R. 2001 Guj. 71. *Taraprawala Vs. Bayer India Ltd.* A.I.R. 1997 S.C. 1846; *M.P. Pollution Board Vs. M.V. Nayudu,* A.I.R. 1997 S.C. 812; *Union Carbide Corpt, Vs. Union of India* A.I.R. 1990 S.C. 273.

26. *Indian Council for Enviro-Legal Action Vs. Union of India,* A.I.R. 1996 S.C. 1446. *Vellore Citizens Welfare Forum. Vs. Union fo India,* A.I.R. 1996 S.C. 2715; Sect. 14 *Residents Welfare Assoc. Vs. State of Delhi.* A.I.R. 1999 S.C. 308; *Rural Litigation and Ent. Kendra Vs. State of U.P.;* A.I.R. 1989 S.C. 594; *M.C. Mehta Vs. Union of India,* A.I.R. 1987 S.C. 965 A.I.R. 1988 S.C. 1037; (1997) 2 SCC 411; A.I.R. 1998 S.C. 2340; A.I.R. 1988 S.C. 2963, A.I.R. 1999 S.C. 291; *L.I.C. of Inida Vs. Manubhai Shah* A.I.R. 1993 S.C. 171.

27. See, for example, *Chhetriya Pradushan Mukti Sangarsh Samiti Vs. State of U.P.,* A.I.R. 1990 S.C. 2060; *Subhash Kumar Vs. State of Bihar.,* A.I.R. 1991 S.C. 420.

28. *M.C. Mehat Vs. Union of India,* A.I.R. 1987 S.C. 965; *Rural Litigation and Env. Kendra Vs. State of U.P.,* A.I.R. 1987 S.C. 359; *Union Carbide Corpt. v. Union of India;* A.I.R. 1990 S.C. 273.

29. *Geogy Mathew Vs. Dy. Comm. of Forest,* A.I.R. 2001 Kant. 404; *Rural Litigation and Ent. Kendra v. State of U.P.,* A.I.R. 1988 S.C. 2187 *Goa Foundation Vs. State of Goa* A.I.R. 2001 Bom. 318.

30. *Union Carbide Corpt Vs. Union of India,* A.I.R. 1992 S.C. 248. Also see *Union Carbide Corpt. Vs. Union of India,* A.I.R. 1990 S.C. 273 where the Supreme Court settled all claims including civil and criminal against 470.

31. *Rural Litigation and Ent. Kendra Vs. State of U.P.,* A.I.R. 1987 S.C. 359.

32. *U.P. Pollution Control Board Vs. Modi Distillery,* A.I.R. 1988 S.C. 1128. *Indian Council for Envior-Legal action Vs. Union of India,* A.I.R. 1996 S.C. 1996 S.C. 1446.

33. *Bangalore Medical Trust Vs. B.S. Mudappa, A.I.R. 1991 S.C. 1902 Union of India Vs. Kamath Holiday Resort* (1996) 1 S.C.C. 774.

34. *Ratlam Municipality Vs. Vardichand,* A.I.R. 1980 S.C. 1622; *B.L. Wadehra Vs. Union of India,* A.I.R. 1996 S.C. 2969.

35. *B.L. Washera Vs. Union of India,* A.I.R. 1996 S.C. 2969; *A.P. Pollution Board Vs. M.V. Nayudu,* A.I.R. 1999 S.C. 812. See a critical comment, C.M. Jariwala, Complex Enviro-Technoloscience Issues: The Judicial Direction, 42 Journal of the Indian Law Institute, 29-39 (2000).

36. *M.C. Mehta Vs. Union of India,* A.I.R. 1987 S.C. 965, 981; *Banwasi Seva Ashram Vs. State of U.P.,* A.I.R. 1992 S.C. 920; *Rural Litigation and Entil. Kendra Vs. State of U.P.,* A.I.R. 1987 S.C. 359, 363.

37. *M.C. Mehta Vs. Union of India,* A.I.R. 1987 S.C. 965; *M.C. Mehta Vs. Union of India,* A.I.R. 1987 S.C. 982.

17

Environment

Judicial Activism and Development of Environmental Jurisprudence

Arvind Kumar Srivastava*

During the last few decades menacing environmental degradation has become the global concern and has aroused the conscience of world polity and it has become the challenge for the present and future as well. "Better late than never" is the apt description of very late awareness of Indian people to the necessity of preservation of Nature and Environment. It is a tragedy that it should happen to a country which has a great spiritual heritage and tradition not only to protect Nature and Environment but to worship the bounties of Nature.

The term environment owes its genesis to a French word "Environ" means 'encircle' and encompasses within it the land, water, flora and fauna, living creatures, forests and everything above the earth. Expression of gratitude to nature as source of sustenance of human life has been our ancient tradition. One of our Rigveda prayers invokes benediction of God for harmony between man, natural resources and energies thus blessings are sought for peace and happiness of all.

Indian philosophy of Vedic period recognizes five natural elements as "panchmahabhut for stavan" meaning worship. These

* Advocate, High Court, Allahabad.

five basic natural elements are water, fire, air, sky and earth. The importance of all the above was highlighted by a great poet "Tulsi Das" : क्षिति जल पावक गगन समिरा : पंच तत्वमिलि बना सरीरा meaning thereby : five basic elements, earth water, fire, sky and air constitute the (human) body.

Since Vedic periods earth and rivers are worshipped as mothers.

"Mata Bhumi Putaro Aham Prithvaiya".

Before putting feet on the earth in Vedic prayer the Rishi seeks forgiveness of mother earth as sustainer of life.

"Vishnupatni Namastubhayam Padsparsh Chhamsav May".

In our Vedic literature every constituent of environment is given an important place. There are some plants and animals whose names were associated with God or Goddesses e.g.- Lion-Durga, Elephant-Ganesh Serpent-Shiva, Peacock-Saraswati etc. Thus the animals and plants were accorded reverence and destroying the trees or killing the animals were considered a sin and it was subjected to various punishments.

Ganga, Godavari, Narmada, Krishna, Kaveri, Bramhaputra and many others are highly venerated rivers of India.Religious Hindus from time immemorial take a round of river Narmada as pilgrimage called "Parikrama". In 'Ishopanishad' there is a verse saying that "a selfish man over-utilizing the resources of Nature to satisfy his own ever increasing needs is a thief because using resources beyond one's needs would result in utilization of resources over which others have a right."

An in-depth study of the literature would further reveal that the ancient Indian law had many commandments on dos' and don'ts.

The natural environment has intrinsic quality of maintaining ecological balance and same is based on environmental management. The mankind has exploited the natural resources and exerted undue pressure on the environment thereby disturbing the equilibrium. The consequences of which are alarming and far reaching. The disturbance in environmental equilibrium polluted the environment which in turn disturbed the ecology.

Some of the serious consequences that emanate form environmental pollution are depletion of Ozone layer, Acid rain, global warming, draught, water, air and noise pollution, loss to bio-diversity, destruction of numerous species of plants and animals etc. There are indirect consequences caused by environmental degradation like allergic skin diseases, development of cataract, cancer, loss of crop yield and loss of hearing.

At this stage it would be relevant to discuss some of the very crucial factors contributing to degradation in environmental quality. Some of these broad factors may be summarized as following:

Water Pollution: Pollution of water is a phenomenon that is characterized by the deterioration of quality of water as a result of different human activities. History reminds us that many civilizations perished or migrated to other locations due to scarcity of water. Water pollution is a global problem and human activities related with water pollution can be attributed to mining, agriculture, stockbreeding, fisheries, forestry, urban development, construction works, industries etc. more or less same all over the world. Like air pollutants, water pollutants produced in one nation may flow into others, creating complex international control problems that may take very long time to solve. Therefore purification, conservation and management of water resources have become important issues for the protection of environment.

Major water pollutants: Following are some major types of water pollutants:

(i) Disease causing agents (Bacteria, viruses and parasites).

(ii) Plant nutrients (Nitrates and phosphates).

(iii) Sediments (Soils, silt and other solids from land erosion).

(iv) Organic & Inorganic chemicals and minerals.

(v) Radioactive substance.

(vi) Oxygen demanding wastes (domestic sewage, animal manure and some industrial wastes).

Sources of Water Pollution: Water pollution can result from natural run off; mixing chemicals in water that percolates thin in the soil and human sources such as agriculture, mining, industry, construction of home and business. There are four main source of water pollution namely (a) Urban Activities; (b) Industrial Activities; (c) Agricultural Activities; and (d) Natural sources such as storm, seepage from ground water, flood etc.

Air Pollution: The most vital component of biosphere is air and without which no living organism can survive. The introduction of large amounts of pollutants into the biosphere causes serious problems for the life on earth. The air pollution is an international problem. According to Sec. 2 of Air (Prevention and Control) of Pollution Act 1981, "Air Pollution" means the presence in the atmosphere of any "air pollutant" which in turn means any solid, liquid or gaseous substance present in the atmosphere in such concentration as may be or tend to be injurious to human beings or other living creatures or plants or property or environment. Therefore it is the duty and legal obligation of all to maintain the quality of air free from pollution.

Sources of Air Pollution: The principal sources of air pollution are as follows:

1. *Domestic Source*– is one of the sources of air pollution caused by the people to meet the basic needs of individual.
2. *Mobile Source*– pollution caused by transport sector has become acute in urban areas. Transport sector emits larger quantities of carbon monoxide, hydrocarbons, oxides of nitrogen and other toxic substances such as lead which are injurious to health.
3. *Industrial Source*– it is a primary source of air pollution. Industrial growth is a must for the development of a nation because of its employment potential, agricultural development and uplifting the standard of living but at the same time its environmental control is our social responsibility. The industrial units engaged in the production and power plants generating electricity are primary source of air pollution.

Influence of Air Pollution

(i) *Green House Effect:* Green House Effects of carbon dioxide are as follows:

(a) The excess concentration of carbon dioxide increases the heat of atmosphere.

(b) Carbon dioxide is transparent for solar radiation i.e. it helps ultraviolet and infrared radiations to reach on the earth.

(c) Carbon dioxide is opaque for thermal radiation and therefore absorbs the heat of the earth. Hence the temperature around the earth is increasing. The above effect of Carbon dioxide is termed as Green House Effect.

(ii) *Acid Rains:* Every source of energy, we use contain sulphur and nitrogen. These two elements when burnt in the presence of atmospheric oxygen get converted into their respective oxides. During rain these highly soluble oxides reach with water vapour and produce acids like sulphuric acid, sulphurous acid, nitric acid and nitrous acid and then return to earth surface with rain water. The said rain is very dangerous to living organism as they can destroy life.

(iii) *Reduced Rainfall*: Rain occurs when cloud droplets coalesce into larger drops or when snow particles are formed in the upper portion of the clouds. Pollution caused by aerosol particles (tiny particles suspended in atmosphere and major components of air pollution) suppressed rain fall by inhibiting coalescence of cloud droplets and by preventing the formation of ice particles in the cloud.

Noise Pollution: Noise pollution is a big problem of our society. Noise is a form of sound. Sound is a mechanical energy from a vibrating surface transmitted by a cyclic series of compression and rare factions of molecules of the material through which it passes. Sound passes through all mediums viz, through air, water, gas, solids but it does not pass through vacuum.

Noise is derived from the Latin word 'nausea'. Encyclopedia Britannica defines noise as "an excessive, offensive, persistent or startling sound." In Acoustics 'noise' is defined as 'any undesired sound'.

Legalistically speaking, noise pollution is the wrongful contamination of the atmosphere due to unreasonable use of sound that it materially and adversely injures the right of an individual to be let alone or interferes in his freedom to lead his life in a manner of his choice. But law does not favour absolute rights, therefore even the right to be let alone is subject to the right of the others to exercise their right like the right to freedom of speech and expression.

The intensity of sound is measured in logarithmic units called decibels. Noise being major health hazard for human beings it is recognized as "Environment Pollution." The Government of India has recently notified the Noise Pollutant (Regulation and control) Rules 2000 for preventing adverse impact on human beings including harmful psychological effects. It provides for ambient air quality standards in respect of noise for different areas/zones as follows:

Area Code	*Category of Area*	*Limits in dB(A) Leg.*	
		Day time 6 am-10 pm	*Night time*
A	Silence Zone	50	40
B	Residential Area	55	45
C	Commercial Area	65	55
D	Industrial Area	75	65

Silence Zone means an area comprising not less than 100 meters around hospitals, educational institutions, religious places and courts. They are to be declared by competent authority that which zone will be silence zone.

Acceptable noise levels as indicated by World Health Organisation (WHO) are at a maximum 45 dB for office & conference room; 40 dB for library, classroom and a living room; for a restaurant and a workshop 60 dB; for a bedroom 25 dB and for hospital wards 35 dB.

Adverse Effects of Noise: Noise is not just an annoyance or a nuisance. It has psychological effects. It constricts the arteries; increases the flow of adrenaline which causes tension; heart beat becomes faster thereby accelerating the rate of cardiac ailments; increase in cholesterol level resulting in permanent constriction of blood vessels; leads to neurosis and nervous breakdown.

Children suffer from neurological reaction that makes them irritable and hyperactive and slows the process of the development of mental faculties. Foetal stress develops in pregnant woman. It disturbs biological organisms and their functioning. It also affects birds and animals. Birds stop laying eggs and breeding. Animals stop mating in the season. More than 155 dB of sound may burn the skin and 198 dB may cause death.

Bio-Medical Waste

There is a growing concern towards proper management of bio-medical waste in the present scenario due to far reaching ill effects resulting from mismanagement of bio-medical waste. The current status of bio-medical waste management is alarming.

Waste is generally defined as "something which is not put into proper usage at a given time. The original definition of waste res derelicta (or the abandoned object) corresponds to the concept of "throw away" culture.

Waste generally is of three kinds:

(i) Liquid waste

(ii) Gaseous waste

(iii) Solid waste

Solid waste is often called the third kind of pollution after air and water pollution. The waste generated in solid state as a result of various human activities and normally discarded as useless or unwanted is termed as solid waste. Handling and disposal of hazardous and non hazardous waste has become a big task due to the growing urban population. Waste can also be categorized according to its origin as domestic waste, industrial waste and hospital waste. Adequate attention has not been paid to collection of hospital waste.

The perils of medical waste first generated attention in the late 1980s, when items such as used syringes washed up on several

East Coast beaches of the USA. This led to the law regulating medical waste. However in India gravity about the issue was realized after 1990s.

Medical wastes are generated through health care facilities wherever medical procedures are conducted. The problem of medical waste has acquired gargantuan proportions and complex dimensions. On an average one hospital bed generates 1 kg of waste per day, out of which 10-15 percent is infectious and the rest is general waste.

Health care waste is a reservoir of potentially harmful micro organisms, which can infect hospital patients, health care workers and general public. Other potential infectious risks include the spread of sometimes resistant micro-organisms from health care establishments into environment.

When bio-medical waste is thrown into the municipal bins, the first victim is the scavenger (Safai Karmachari) and second is rag picker. Rag pickers who rummage through this waste in search of disposables are at great risk. They are exposed to the waste through any open wounds; inhalation of air borne pathogens and pricks of sharps. Bio medical waste can also cause injuries for example radiation burns or sharp-inflicted injuries, poisoning and pollution whether through the release of pharmaceutical products, in particular antibiotics and drugs through the waste water or by toxic elements or compounds such as mercury or dioxins. Because of this improper handling, treatment and disposable of the bio-medical waste chances of the spreading of diseases such as HIV, Hepatitis and ocular, genital and skin infections are more.

e-Waste

With the advent of electronic era a new type of pollution is taking place around the globe specially India. Nowadays each and every person as well as every house is having electronic goods or instruments such as TV, Computers, Mobile Phone, VCR, DVD, MP3, electronic music system, CD. Even the modern kitchen is being fitted with electronic chimney. These electronic goods have hazardous components or parts such as chips, hard disk, picture tube, ram, motherboard, floppy etc. There is a growing concern towards proper management of e-waste is the present situation. Handling and disposal of e-waste is becoming a big task due to growing urban population which is using electronic goods at a increasing pace.

India's capital is emerging as the world's capital for e-waste recycling, a hazardous activity taking place without any regulations and restrictions. Conservative estimates put the annual volume of e-waste-the junk produced by discarded electronic equipments in Delhi at around 12,000 metric tones, but experts say the actual figure is even higher. Delhi alone gets around 70 per cent of electronic waste generated in developed world.

e-waste are generally generated through recycling units where personal computers and TV sets are dismantled for reuse of there components, parts and materials.

e-waste has not attracted the attention of the authorities even common man is not concerned with the spread of e-waste. There is a growing need for proper legislation in this field. A detailed study in required for the harmful effects of the e-waste in the environment.

Judicial Activism

Judicial Activism does not carry any statutory definition. Broadly, it is understood as connoting that function of judiciary which represents its active role in promoting justice.

In order to be able to understand its meaning it is important to first understand the meaning of the word 'Activism' which according to Chamber's 20th Century Dictionary means a policy of vigorous action of a philosophy or a creative will.

In Webster's New Twentieth Centaury Dictionary (unabridged) the word 'Activism' has been stated to mean 'the doctrine or policy of being active or doing things with decisions.'

Judicial Activism on the basis of above two meanings would mean an 'Activism' by taking recourse to judicial process leading to judicial pronouncements on different intricate issues whereby new approach towards legal philosophy is sought to be developed by recourse to dynamism in discharge of judicial functions.

Judicial Activism is nothing more than a description of performance of duty consistently with the needs of the time, by the judiciary and the Judges who are active; and within the parameters of the Constitution and the laws. While doing so they exercise to brim their constitutional jurisdiction in larger public interest, and with innovation, as they are convinced that nothing lesser will serve the public purpose. Judicial Activism is recognized by the Courts of their constitutional obligation in the circumstances where failure to act would be betraying the people's faith in judicial system of a constitutional democracy. Judicial Activism in essence and essentially aims at upholding the Rule of Law impregnated with truly, democratic values.

The concept of Judicial Activism is not one of the recent past it was born in 1803 when Chief Justice Marshall, the great judge of the American Supreme Court decided Marbusy vs Maddison (1803) 2L Ed.60. He observed that the Constitution was the fundamental and paramount law of the nation and "it is for the court to say what the law is". He concluded that the particular phraseology of the Constitution of the United States confirms and strengthens the principle supposed to be essential to all written Constitutions: that a law repugnant to the Constitution is void and that the courts as well as other departments are bound by that instrument. If there was conflict between a law made by the Congress and the provisions in the Constitution, it was the duty of the court to enforce the Constitution and ignore the law. The twin concepts of judicial review and judicial activism were thus born. Article 142 of the Indian Constitution is a unique provision of the Indian Constitution, which probably is not there in any other Constitution of the world, conferring unfettered powers on the Supreme Court to pass any decree, order or direction, which the interest of justice demands and/or which may be necessary for doing complete justice.

Article 141 of the Constitution states that 'the law declared by Supreme Court shall be binding on all courts within the territory of India.

Both these Articles make the way for Judicial Activism. A few factors have contributed to the emergence of the trend of judicial activism. First, the judiciary as an institution has been able to command the respect and confidence of the people especially in the years which have followed the infamous emergency. People saw their only saviour in the judiciary. Secondly, the power of the Court to punish for its contempt, a powerful weapon though rarely used, deters the persons ordained from disobedience. Thirdly, the fractured poll verdicts have created an atmosphere of political uncertainty. The Legislature too becomes weak. The Government is not much inclined to take decisions which would invite the displeasure of the people who are their voters. Many a times, the judicial activism is invited by the Government and the executive as a friendly measure and as a shield for themselves. In the corridors of power, it is very common to listen – "the matter is pending in the Court, await the decision" or "why not have the matter decided by the Court". Quite often the Government, and in particular the executive believes in shifting the responsibility at times passing on the buck, by inviting Court's decision rather than taking a decision by themselves. Mostly this happens in such areas as are relatable to the issues which are politically, religiously or socially sensitive. For example take the case of Taj corridors, shifting of abattoirs, demolition of encroachments, and pollution of Ganga Water and so on.

Judicial Activism in Environmental Law: In India, the solution of problems on environmental issues may be found in four branches of law: crime, fort, statutory regulations and the Constitution of India.

Constitutional Provisions: When the Constitution was framed and came into operation in 1950 we find no mention of Environment in either Part III relating to "Fundamental Rights" or Part IV relating to Directive Principles of State Policy. It is only in the year 1976 that Article 48A was introduced as a new environmental provision in Directive Principles of State Policy which says "The State shall endeavour to protect and improve the environment and to safeguard the forests and wild life of the country.

Article 37 provides that although the directive principles "shall not be enforceable by any court", they are nevertheless fundamental in the governance of the country and it shall be the duty of the State to apply these principles in making laws.

However Article 37 makes Article 48A not to be enforced by any court. But the Rajasthan High Court in L. K. Koolwal vs State of Rajasthan, AIR 1988 Raj 2 has put a new shade in this so called provision by evolving a fundamental right to live in clean environment.

A new provision on fundamental duties by Article 51A was introduced by 42nd Amendment in 1976, which says in clause (g), "it shall be the duty of every citizen to protect and improve the natural environment including forests, lakes, rivers and wild life, and to have compassion for living creatures. There is no mention of their enforcement by writ (Article 32 and Article 226) or other legal provisions.

Statutory Regulations: There are many enactments for preserving and protecting the environment in its existing form and control the pollution thereof. Some of the important legislation are mentioned as follows:

Fair Way Act 1881, The Police Act 1861, Northern Canal and Drainage Act 1873, Oriental Gas Company Act 1857, Poison Act 1919, Indian Boilers Act 1923, The Sailors Act 1923, Indian Forest Act 1927, The Air Craft Act 1934, Factories Act 1948, The Industries (Development and Regulation) Act 1951, Mines and Minerals (Regulation and Development) Act 1957, The Insecticides Act 1968, The Wildlife Protection Act 1972, The Code of Criminal Procedure 1973, The Water (Prevention and Control of Pollution) Act 1974 and rules there under. The Water Pollution Cess Act 1977. The Forest Conservation Act 1980, The Air (Prevention and Control of Pollution) Act 1981, The Environment Protection Act 1986, Motor Vehicles Act 1989. Noise Pollution (Regulation and Control) Rules 2000 under The Environment (Protection Act) 1986.

Having so much of Acts which came into existence from time to time to need of prevention of pollution from the environment. The environmental pollution causing nuisance is repressible for an action under section 133 Cr.P.C. The nuisance coming within the purview of civil wrong in repressible by an action through and individual or by an action in the representative capacity under section 91 C.P.C.

Seeing the gravity of pollution caused by Bio-Medical waste The Ministry of Environment and Forests promulgated Bio-Medical Rules in 1998 in exercise of the powers conferred by section 6, 8 and 25 of The Environment Protection Act 1986. Bio-Medical Rules is the first of its kind of national law in the entire South East Asian region in relation to bio-medical waste management.

The general provisions of The Indian Penal Code 1860 can also be applied for the management of bio-medical waste management. A polluter of the environment can also be punished if he does an act which causes any common injury danger of annoyance to the public or to the people in general then the act may be treated as public nuisance as defined under section 268 and the offender may be punished under sections 290 and 291 of the Code. There are also penal provisions for a situation which either causes, or destroys, or diminishes the value or utility of any property, or affects the property injuriously as provided under sections 426, 431 and 432 of the Code. This otherwise means if any person who generates, collects, receives, stores, transports, treats, disposes or handles bio-medical wastes in any form shall be treated as contravener of the above penal provisions.

With the changes in times, and to cope with new challenges, judiciary has innovated new methodologies. Public Interest Litigation (PIL) is a unique device wherein shedding the shackles of locus standi, the judiciary entertains grievances of public nature under Articles 32 and 226 of the Constitution. Environmental consciousness has also rendered support to the demand of judicial activism in the field of environment. Art 142 (1) of the constitution has become a great source of judicial activism.

The polluter pays principle refers to requirements that cost of pollution be borne by the person or persons responsible for causing pollution and the consequential cost.

The precautionary principle—the meaning of precautionary principle is not clearly defined but it conveys that where the exploitation of natural resources for community use pose threats of serious or irreversible damage, lack of scientific certainty shall not be a reason for not undertaking cost effect measures to check environmental degradation.

Before the RIO treaty of 1992 and without reference to such concepts as 'Sustainable Development', 'Precautionary Principle' and 'Polluter Pays Principle' can be traced to the Supreme Court decision in Ratlam Municipality vs Virdi Chand AIR 1980 SC 1622. In this case Supreme Court has recognized the importance of pollution free environment and gave it the status of human right. The Supreme Court further stated that responsible municipal council constituted for the precise purpose of preserving public health and providing better finances can not run away from its principal duty by pleading financial inability. Expanding the scope of Article 21, the Supreme Court in M.C.Mehta vs Union of India (1997) 3 SCC 715 further held that the "precautionary principle" requires the state to anticipate, prevent and attack the causes of environmental degradation.

In Subhash Kumar vs State of Bihar AIR 1991 SC 420, and Chhetriya Pradushan Mukti Sangharsh Samiti AIR 1980 SC 1622 the Supreme Court has treated right to pollution free environment as integral part of right to life guaranteed by Article 21 of the Constitution.

The Bhopal accident happened two year ago before the Supreme Court evolved the rule of Absolute Liability. The Court in this case was faced with the problem of making quick decisions relating to compensation of victims. To avoid the problem of parties the Parliament passed the Bhopal Gas Leak Disaster (processing of claims) Act 1985. In Charan Lal Sahu vs Union of India AIR 1990 SC 1480 the validity of Bhopal Gas Disaster (processing of claims) Act 1985 was challenged. Upholding the validity of the Bhopal Act 1985 Justice K. N. Singh made valuable remarks. The Judge observed that in context of India's national dimensions of human right to life, liberty, pollution free air and water is guaranteed by the Constitution under Articles 21, 48 and 51 (g), and therefore it is the duly of the State to take effective steps to protect the guaranteed constitutional rights.

Chief Justice R.N. Mishra expressed judicial concern in the Bhopal Gas Leak industrial accident as 'Judges of this court are men and their hearts also bleed when clamities like Bhopal Gas incident occur'.

The Bhopal Case had resolved some important aspects like the 'Industrial Disaster Fund'. It also initiated legislative activism and the Public Liability Insurance Act 1991 came into being to provide insurance of public liability and immediate relief to victim of an accident which occurred while handling hazardous substance.

By interpreting Article 21 of Citizens' Fundamental Right to Life' the Supreme Court expanded its meaning to read several basic human rights into it. This jurisprudence fully developed with the decision of Supreme Court in the case of Fancis Coralic Mullin vs Administrator Union Territory of Delhi (AIR 1981 746) which said the right to live with dignity. In M. C. Mehta vs Union of India AIR 1987 SC 1084 and Bandhua Mukti Morcha vs Union of India (AIR 1984 SC 802) held that right to life include right to healthy environment. The Supreme Court further expanded and said that right to healthy environment itself include the pollution free water and air (B. L. Wadhera vs Union of Indian AIR 1996 SC 2969 and Indian Council for Enviro-Legal Action vs UOI AIR 1996 SC 1446) and protection against hazardous industries (Vellore Citizens Welfare Forum vs Union of India AIR 1996 SC 2715) and further held that the precautionary principle and polluter pays' principle are part of the environmental law of this country.

In M.K. Sharma vs Bharat Electronics Ltd AIR 1987 SC 1792, the workers of public sector undertaking filed a case claiming compensation for being exposed to the ill effects of X-ray radiation, the Supreme Court issued direction as to check and safeguards to be adopted to guard against radiation.

The Supreme Court said that if an industry is established without requisite permission and in blatant disregard of law to the detriment of citizens' right to life, the Court can interfere to protect right to life and in such case rule of absolute liability applies (Indian Council for Enviro-Legal Action vs Union of India AIR 1996 SC 1446).

Development of Environmental Jurisprudence can be witnessed in the case of Banwasi Seva Ashram vs State of U. P. (1986) 4 SCC 753 here Supreme Court gave directions regarding declaring disputed areas as reserved forests.

The Supreme Court has enunciated the doctrine of "Public Trust" based on the legal theory of the ancient Roman Empire. The idea of this theory was that certain common properties such as rivers, seashores, forests and the air were held by the Government in trusteeship for the free and unimpeded use of the general public. The resources like, air, sea, waters and the forests have such a great importance to the people as a whole, that it would be totally unjustified to make them a subject of private ownership. The concept "environment" bears a very close relationship to this doctrine. The doctrine enjoins upon the resources for the enjoyment of the general public, rather to permit their use for private ownership or commercial purposes. It was thus held that the State Government Committed breach of public trust by leasing the ecologically fragile land to the motel management [M. C. Mehta vs Kamal Nath (1997) 1 SCC 388].

Supreme Court opined that every attempt should be made to preserve the fragile ecology of the forest area and to protect the Tiger Reserve and the right of tribals in the State of M. P. [Animal and Environment Legal Defence Fund vs Union of India (1997) 3 SCC 549]

The compensatory jurisprudence to combat environmental hazards was extended by imparting remedies for occupational disease that destroy workers engaged in hazardous industries in Consumer Education and Research Centre vs Union of India AIR 1995 SC 922 by the Supreme Court.

In Murli S. Deora vs Union of India AIR 2002 SC 40 the Supreme Court banned the smoking in public place and said that in any case, there is no reason to compel non-smokers to be helpless victim of air pollution. The Supreme Court held in Pradeep Sharma (P) Ltd vs Union of India AIR 1997 SC 2648 and Lakshmikant vs. Union of India (1997) 4 SCC 739's case that total prohibition can be imposed against manufacture of drugs or preparations which are injurious to health.

After the decision in Maneka Gandhi AIR 1997 SC 597 if unnamed right is an integral part of named fundamental right it acquires the status of a fundamental right looked at from this angle, loudspeaker is an effective tool to exercise the freedom of speech and expression. Therefore it was urged that right to use loudspeaker is a fundamental right. The Court rejected the contention and the rights of an unwilling listener were upheld with an observation "apart from the right to be alone—freedom from aural aggression. Article 21 guarantees freedom from tormenting sounds i.e. noise pollution.

Landmark decision made by the Supreme Court in Church of God (full Gospel) in Indian vs KKRMC Welfare Association AIR 2000 SC 2773 where the Court denied the right of the Church of God to use amplifier and other noise making devices for prayer and observed that the noise rule are required to be enforced and further said that it is high time when they should be implemented.

The Bio-medical wastes attracted the attention of the Supreme Court as early as in the year 1994 when a writ petition was filed under Article 32 of the constitution against the concerned authorities to provide clear and healthy environment. Issue of improper hospital waste management was the crux of the case in B. L. Wadhera vs Union of India AIR 1996 SC 2969. In this case the Supreme Court issued 14 direction of which 4 were related to bio-medical waste.

1. All hospitals with 50 beds and above should install incinerators or any of effective alternative method under their own administrative control.
2. AIIMS, New Delhi through its director install sufficient number of incinerators or any equally effective alternative to dispose of hospital waste.
3. MCD and NDMC issue notices to all the private hospitals/nursing homes in Delhi to make their own arrangements for the disposal of their garbage and hospital waste. In case these hospitals are permitted to use facilities provided by the MDC and NDMC then they may be asked to pay suitable charges for the service rendered in accordance with Law.
4. The Central Pollution Control Board and the State Pollution Control Boards should regularly send its inspection teams in different areas to ascertain the collection, transportation and disposal.

Later in Almitra Patel vs Union of India AIR 2000 SC 1256 the attention of The Supreme Court was drawn to the 14 directions in B. L. Wadhera case. The court showed its deep anguish regarding the non-observance of directions and said in these words "it is indeed unfortunate that till today the said directions have hot been complied with."

Doctrines and guidelines set by the Apex Court shows Judicial Activism is not an aberration. It is an essential aspect of the dynamics of a Constitutional Court. It is a counter majoritarian check on democracy. Judicial activism does not however mean governance by the judiciary. It also must function within the limits of judicial process. To preserve the sanctity and credibility of the judicial process and to over come of judicial activism, it is necessary to practice self restraint while innovating new tools. It is no doubt true that despite activist role of judiciary, a better mechanism needs to be developed for solving environmental conflicts and issues as they are so precious to human life. Judiciary is playing its humble role. Apex Court is final but not infallible.

The contribution of Indian Judiciary towards the development of environmental jurisprudence has been very good. The judiciary has taken active part in rendering guidance towards the preservation of environment. Our judiciary has assessed the concept of Environment Impact Assessment and every other jurisprudential aspect of environment. The proposal for the creation of Environmental Tribunal imparting of environmental education to bring social awareness on environment were other innovations of the Indian Judiciary. The enforcement machinery and administrative authorities also were made to work efficiently in many cases like shelving of 'Silent Valley' project in Kerala, stopping of limestone quarries in Doon Valley, control of vehicular pollution in Delhi, closer of polluting units in Agra to save Taj Mahal, relocation of polluting industries out of residential areas of cities and Taj Trapezium case (M. C. Mehta vs Union of India AIR 1997 SC 734). Therefore the contribution of judiciary in development of environmental jurisprudence is a creative work, which has to be acknowledged by us, and we should try to incorporate it in our legislative, where it fails to do so.

The judicial pronouncement reflects the value of environment protection. The Supreme Court has always upheld the preciousness of environment not by choice but as being a need of the situation. Judicial verdicts have given appropriate dimension to development of environmental jurisprudence.

REFERENCES

1. Justice D. M. Dharmadhikari, Supreme Court of India: "Environment-Problems and Solution" All India Reporter 2003.
2. Raghvendra Kumar: "Environment Legal and Juridical Scene" All India Reporter 2002.
3. Justice R. C. Lahoti, Chief Justice of India: "Judicial Activism—Constitutional Obligation of the Courts," All India Reporter 2003.
4. Dr. S.A.K. Azad: "Water Pollution and its Legal Control in India," All India Reporter 2001.
5. Dr. S.A.K. Azad: "Air Pollution and its Legal Control in India," All India Reporter 2001.
6. Dr. Veena Madhav Tonapi : "Noise Pollution : Are the Laws Audible?" All India Reporter 2005.
7. Dr. S.R. Sharma : "Noise Pollution and Public Interest Litigation," All India Reporter 2005.
8. Kiran M. Nadagoudar : "The Law Relating to Management of Bio-Medical Waste : A Critical Analysis," All India Reporter 2003.
9. The Times of India—Delhi Edition Sept. 2006.
10. C.M. Jariwala: "Indian Environmental Law: The Finished and Unfinished Agenda".
11. Madhumita Dhar Sarkar: "Contribution of Indian Judiciary Towards the Development of Environmental Jurisprudence," All India Reporter 2005.

18

Environmental Provisions in the Indian Constitution

Dr. Tarun Kumar*

After the Stockholm Conference of 1972, a new phase of debate to protect environment began and since then it has become a world wide issue. The Stockholm Conference has burdened upon the governments and people to exert common efforts for the preservation and improvement of the human environment for the benefit of all mankind for their posterity.

The Stockholm Declaration (1972) on Human Environment proclaimed that:

"Man is both 'creator' and 'moulder' of his environment which gives him physical sustenance and affords him the opportunity for intellectual, moral, social and spiritual growth. In the long evolution of the human race on this planet a stage has been reached when through the rapid acceleration on sciences and technology, man has acquired the power to transform his environment in countless ways and on an unprecedented scale. Both aspects of man's environment, the natural and the man-made, is essential for his well being and to the enjoyment of basic "human rights", even the right to life itself."

* **Senior Lecturer in Political Science, L.B.S.P.G. College, Mughalsarai, Chandauli.**

U.N. Proclamation also adopted that all governments of the world realize human environment as a major issue. Apart from governmental initiatives man himself must be committed to build a better environment by using his knowledge in collaboration with nature. Thus the principles of the Stockholm Declaration (1972) are known as Magna-Charta on human environment as they provide a basic code of environmental conduct. It emphasized the urgent need of intensifying the efforts at the global, regional and national level to protect and improve the human environment.

The Stockholm Conference placed environment as an explicit subject of international law and policy on the global political agenda. United Nations was named as the principal formulator of an international law on the environment and international environmental diplomacy moved towards universalization of international agreements on global environment.

In short, the Stockholm Conference has ushered in an era of global environmental consciousness and activity and the institutional frame work provided by the UNEP, IGOS and INGOS has been instrumental in bringing about varied policy responses at national regional and global levels in relation to current and emergent environmental issues. The conference adopted the motto "Only one Earth" for the entire humanity. The conference declared June 5th as the world Environment Day.

The Indian environmental policy is unavoidably linked with the international policies and global situation. Originally the Constitution of India did not contain provision having a direct bearing on environment but some of the provisions do make tangential reference about environment particularly in the provisions of Directive Principles of State Policy. After 42nd Amendment of the Constitution environmental provisions contain a clear mandate to the citizens of India to protect and improve the environment. In India there is also an independent Department of Environment, Forests and Wildlife. Various enactments have entered into the statute book to safeguard environment and proper action are being pursued but the problems of environmental degradation continued.

After independence, India followed the ideal of socialism and for the first time this ideal was given a constitutional recognition by the 42nd Constitutional Amendment Act, 1976. For the protection of environment serious efforts began with the establishment of National Committee on Environmental Planning and Co-ordination (NCEPC) by the department of science and Technology in February, 1972.

There are several enactments about the protection of environment, the most important being the Factories Act, 1948, the Insecticides Act, 1968 and Motor Vehicles Act, 1939. After Stockholm Conference, 1972 The Parliament of India has enacted a number of laws, directly relating to pollution control and environment protection, e.g., the Water (Prevention and Control) Act, 1981, the Forest (Conservation) Act, 1980, the Wild Life (Protection) Amendment Act, 1986 and the Environment (Protection) Act, 1986. In addition to these, several state governments have enacted special laws which can be evoked for protecting and improving environment.

Fundamental Rights and Fundamental Duties including Directive Principles contain a clear mandate to the Citizens of India to protect and improve the Environment. No doubt, India is one of the few countries which have provided constitutional safeguards to preservation of the environment.

The Fundamental Rights are not directly concerned with environment but its soul is linked with environment which came in light by judicial activism. The judiciary has exhibited and enlightened judicial creativity and foresight whenever it has opportunity, to decide issues relating to environment. The Supreme Court opened the door for realization of fundamental duties with regards to preservation of environmental to both Government and Citizens. The Supreme Court also realized that failure or delay on the part of administration to attend to the urgent task of enforcing anti-pollution and anti-deforestation to laws and regulatory measures in the national interest was considered to be the violation of the Fundamental Rights of people conferred by Articles 14 and 21 of the Constitution of India.

The Supreme Court interpreted the right to life and personal liberty to include the right to a whole some environment. Thus the right to human environment has been recognized as a part of the right to live guaranteed under the constitution of India. For example, In the "Oleum gas Leakage Case" of New Delhi based Sri Ram Fertilizers, the Supreme Court realized that "the negligence of the management in the maintenance and operation of the caustic chlorine plant and the indifference shown by the management in installing proper safety devices and instruments and taking proper and adequate measures for ensuring safety of the workmen and the people living in the vicinity can not be overlooked". That is why, an integrated approach to deal with the environmental pollution problems, the Environmental (Protection) Act was passed in 1986 in the Parliament of India.

Article 14 of the Constitution enunciates the fundamental right of every person not to be denied, "Equality before the law" the "equal protection of laws" within the territory of India. It embodies the principle contained in the Universal Declaration of Human Rights that "All are equal before the law and are entitled without discrimination to equal protection of the law."

The guiding principle underlying Article 14 is that all persons and things similarly circumstanced shall be treated alike both in privileges conferred and liabilities imposed. Laws should be applied to all in the same condition. (State of West Bengal Vs Anwar Ali Sarkar AIR 1952 SC 75, and Bennett Coleman & Co. Vs Union of Indian, AIR 1973 SC 106).

Justice Bhagwati became more forthright in his approach to Article 14 which received unanimous approval of a constitution Bench of the Court in the following words:-

"It must ... now be taken to be well settled that what article 14 strikes at is arbitrariness because an action that is arbitrary, must necessarily negation of equality. The doctrine of classification which is evolved by the Court is not paraphrase of articles 14 nor is it the objective and end of that article.

Thus, the new approach propounded by the Supreme Court has widened the Scope of application of article 14 and now one need not allege any discrimination vis-à-vis others (A. L. Kalra V. Project and Equipment Corporation, 1984, 3SCC316, 328). It is clear that arbitrary or unreasonable actions would be per se discriminatory and violative of article 14.

The Supreme Court through its wisdom and activism has expanded Article 21 and it is hoped that a day would come soon when it may declare environmental pollution dangerous to life as violation of Article 21 of the Constitution. For Example, In the case of Maneka Gandhi Vs Union of India, (AIR 1978 SC 597), a wider and liberal interpretation was given by the Supreme Court as the right to life in order to include the right to live with human dignity and decency. The Supreme Court in fact over-ruled the Gopalan case expressing the view that "the attempt of the Court should be to expand the reach and ambit of the Fundamental Rights rather than to attenuate their meaning and context by a process of Judicial Construction." It held that the right to 'Live' is not merely confined to physical existence but it includes within its ambit the right to live with human dignity. Elaborating this view in Francis Coralic Vs Union Territory of Delhi, the court said that the right to live is not restricted to mere animal existence. (AIR 1981 SC. 746).

In the case popularly known as the 'Pavement Dwellers' case, the Supreme Court observed that the word "Life" in article 21 included the "right to livelihood." It said that if the right to livelihood was treated as a part of the constitutional right to life, the easiest way of depriving a person of his right to life would be to deprive him means of livelihood (Olga Tellis Vs. Bombay Municipal Corporation, AIR 1986 SC. 180).

Thus, it is quite clear that the fulfillment of human fundamental needs is dependent on many elements of the environment air to breath, water to drink, food to eat and shelter for protection. The quality of the environment is therefore, directly related to the full enjoyment of the right to life. As the constitutional Court of India found:-

"The right to live is a fundamental right under article 21 of the Constitution and it includes the right of enjoyment of pollution-free water and air for full enjoyment of life. If any thing endangers or impairs that quality of life in derogation of laws, a citizen has the right to have recourse to article 32 of the Constitution for removing the pollution of water or air which may be detrimental to quality of life."

In another substance, the Kerala High Court observed in the case of 'Attakoya Thangal Vs. Union of India' that the right to sweet water and the right to free air, are attributes of the right to life for these are basic elements which sustain life itself."

In fine, the fundamental right to life of people in article 21 of the Constitution has preferred over right to life of individual because life, health and ecology have greater support by Justice Singh's concluding observation in Ganga Pollution case. The Supreme Court in this case observed:-

"The petition in the case before us is no doubt not a riparian owner. He is a person interested in protecting the lives of the people who make use of the water following in the river Ganga and his right to maintain the petition can not be disputed".

Party IV of the Constitution of India deals with the Directive Principles of State Policy inspired by the Irish precedent, are a unique feature of our Constitution. Most of the socio-economic rights of the people have been included under this head. Though it is a directive principle and thus not enforceable through Courts, Yet its directive character does not dilute its significance and the obligation of the state to protect and improve environment. The Directive Principles have increasingly assumed greater relevance and importance not only for the legislatures but also in the eyes of the Court.

Article 47 of the Constitution of India says: "The State shall regard the raising of the level of nutrition and the standard of living of its people and the improvement of public health as among its primary duties and, in particular, the state shall endeavour to bring about prohibition of the Consumption except for medicinal purposes of intoxicating drinks and drugs which are injurious to health." This article provides for the improvement of the public health as one of the primary duties of the state and implication of the protection and improvement of the environment.

The 42nd Amendment Act, 1976 added a clause 48A, even 48 prohibited cow slaughter. Article 48A says. "The State shall endeavour to protect and improve the environment and to safeguard the forest and wild life of the Country." This article is an obligation of the state to protect and improve the environment. To make it effective few insertions are made in the legislative lists enshrined in the Seventh Schedule of the Constitution.

The Indian Constitution is one of those few Constitutions of the world which has provided a constitutional duty of the Citizens to protect and improve the environment. The 42nd Amendment Act, 1976 has incorporated into the constitution, a new chapter on Fundamental Duties Part IVA. In this chapter particularly Article 51A (g) shows a deep concern of the state to protect and preserve the natural environment. This Article provides : "It shall be the duty of every citizens of India to protect and improve the natural environment including forest, lakes, rivers and wild life and to have compassion for living creatures."

Thus it is the constitutional obligation of the state and citizens of India to protect and improve healthy environment. No doubt, in the absence of healthy and clean environment, existence of human life is in danger. Therefore, all citizens should be alert for healthy environment. This view was accepted by Justice Ranganath Mishra, former Chief Justice of India. In his words:-

"Preservation of the environment and keeping the ecological balance unaffected is a task which not only Government but also every citizen must undertake. It is a social obligation."

The Review of the Constitution of India

The Constitution Review Commission headed by Justice Sri M. N. Venkatachalliah, former chief justice of India was constituted by a Resolution of the Ministry of Law, justice and Company Affairs on February 22, 2000 which submitted its report on 31st March, 2002.

The Review Commission has mentioned some Articles related to Environment. In regard to fundamental rights, Review No. 16 After the proposed Article 30-C the following article may be added as Article 30-D:- 30-D, Right to safe drinking water, prevention of pollution, conservation of ecology are sustainable developments."

Every person shall have the right: (a) to safe drinking water; (b) to an environment that is not harmful to one's health or well being; and (c) to have the environment protected, for the benefit of present and future generations so as to:

(i) prevent pollution and ecological degradation;

(ii) promote conservation; and

(iii) secure ecologically sustainable development and use of natural resources while promoting justifiable economic and social development.

Thus, Article 30-D ensures right to safe drinking water, prevention of pollution, conservation of ecology and sustainable development as fundamental right.

Further the Commission recommended that the part IV of the Constitution should be amended to read as 'Directive Principles of State Policy and Action.

The Review Commission also suggested a new Article 47A "Control of population-the state shall endeavour to secure control of population by means of education and implementation of small family norms."

Judicial Activism and Human Rights

Human Rights are those rights which are inherent in nature and without which no one can survive. These rights are available to each and every human being against the state or other public authority. The Stockholm Declaration on the Human Environment of 1972 first pronounced on the inter-relationship between the enjoyment of human rights and the quality of the environment. Since then, the conception of this issue has taken many forms. One has been to add a "right to environment" to the human rights catalogue. The Stockholm Declaration fell short of proclaiming such a right, it emphasized that the full enjoyment of human rights required the environment.

The African Charter on Human and Peoples Rights states that "all people have the right to a general satisfactory environment favourable to their development. Similarly American Convention on Human Rights grants an individual human "right to live in a healthy environment." The Rio Declaration on Environment and Development may reflect the persistence of the doctrinal controversy by proclaiming merely that "Human beings are entitled to a healthy and productive life in harmony with nature."

At present, all of us have to think about violations of the right to life, the right to the peaceful enjoyment of one's positions. Clean environment has now become essential to human life.

Therefore, on account of multidimensional effects of pollution, right to healthy environment is universally accepted as human right.

However, the judiciary of India has come out as the guardian and the custodian of the Indian Environment. It has played significant role in interpreting and expanding the scope of fundamental rights as "Right to life". The right to life as enshrined in Article 21 of the Indian Constitution has been expanded to human rights. This interpretation has been done by Supreme Court in "*Olean Gas Leakage Case*" of New Delhi based Sri Ram Fertilizers, the Supreme Court realized that right to life is incomplete without right to healthy environment. The Supreme Court has observed that, "This is a question which needs serious attention of the government of India to take the necessary steps at the earliest because the problem of danger to health and well being of the community on account of Chemical and other hazardous industries has become a pressing problem in modern industrial Society."

In a developing country like India facing with problems of poverty, illiteracy and Socio-economic inequalities, judiciary can not be concerned merely with the isolationist character of human rights, but this norm is first time broken and environmental pollution was controlled through the Supreme Court in the Rural Litigation and Entitlement Kendra and M. C. Mehta cases bear testimony to the fact that the traditional rule of laws did not stand in the way in cases where environmental questions were raised and the Supreme Court interfered and gave direction after direction to the Government for taking environmental protection measure in the interest of the general public.

In another substance, the ecosystem of Doon Valley in Mussoorie Hill of Himalaya was being destroyed by illegal and unauthorized mining activities for dolomites. Therefore, the

Supreme Court stopping all fresh quarries observed that, "We are not oblivious of the fact that natural resources have got to be tapped for the purpose of social development but one can not forget at the same time that tapping of resources have to be done with requisite attention and care so that ecology and environment may not be affected in any serious way".

(In Rural Litigation and Entitlement Kendra Vs. State of Uttar Pradesh—AIR- 1987 SC)

The river Ganga and Yamuna of North India, remains badly polluted by the Industrial effluents. In this regards, the Supreme Court ordered in 1995 under the provisions of EPA, 1986 and asked for shifting of all those polluting industries settled on the bank of river Ganges in Kanpur and Hooghly in Kolkata and those on the bank of Yamuna in Delhi.

The Supreme Court has acted as a saviour of environmental rights. For example, on 28th July, 1998 the Supreme Court has directed to D.T.C. and Private operators of Delhi to convert 15000 buses into C. N. G. fuel till March 31, 2001 which was further extended to September 30, 2001.

Here, it can be said that C. N. G. fuel is free from pollution.

The Supreme Court on July 21, 2003 issued notice to the Centre and the U. P. Government on a petition seeking a direction to urgently clean up the 80 – odd bathing ghats on the banks of Ganga River in Varanasi where thousands of Pilgrims and tourists from across the world take a dip every day.

Lawyer Ashok Kumar Thakur said, "The rich heritage and our composite culture and holy Ganga and its natural environment, with the Ghats since time immemorial, are required to be preserved and protected in the public interest."

The Supreme Court also on July 25, 2003 asked a central technical committee to assess the damage potential of construction material swept away by monsoon to the Taj Mahal in Agra. This order was passed by a Bench, comprising Justice M. B. Shah and Justice A. R. Lakshmanan, after hearing court commissioner Krishan Mahajan's plea for inspection of the site by experts to assess the danger posed to the Taj Mahal and the Agra Fort from the rubble washed away by rain water from the construction site of the corridor project.

Thus, on the basis of some cases (by judicial activism) the Supreme Court has acted as a protector of human rights and environment.

Despite several constitutional provisions, legislation and laws, environmental scenario is deteriorating day by day. The awareness among the people is still lacking so the need of the hour is to change the mindset of the people so as to induce them to act according to Dharma. In ancient India, the environmental protection was the Dharma of each and every individual, our fundamental duty is to protect and improve the natural environment and to safeguard the forest, lakes, rivers, wildlife and to have compassion for living creatures.

In short, we can suggest that new dimensions to 'National Security' can not be only introducing new weaponry systems among our armed forces but priority action is called for restriction, conservation and development of our environment and ecosystem. This can be achieved by stabilizing the population growth, integrated, control of pollution in water and air. Environmental education and awareness among the people and effective environmental laws are most necessary for the protection of environment. Amendment in the Indian Constitution should be made in this regard. For this Constitutional Review Commission's recommendation will be more effective mainly in the field of the Fundamental Rights and Directive Principles.

I conclude, in the words of B.D. Nagchandhuri and S. Bhatt that the environmental policy in India would take an integrative view of changes in environment and overall psycho-social development of man. An action-oriented scientific response is needed to stop deterioration of environment. It is often possible to improve environment without unduly disturbing the ecological balance. For this purpose, manmade laws should be carefully analyzed so that there is no conflict with nature's laws.

REFERENCES

1. B. D. Nagchandhuri and S. Bhatt, *The Global Environment Movement: A New Hope for Mankind,* Sterling Publishers Private Limited, New Delhi, 1987.
2. N. R. Inamdar, "Environmentalist Movement and Development", Deepender Basu (ed.), *Environment and Ecology : The Global Challenge,* Prinwell Publisher, Jaipur, 1995.
3. Rajiv Kumar Sinha and Sunderlal Bahuguna (ed.) *Environmental Crisis and Human at Risk,* Ina Shree Publishers, Jaipur (India), 1999.
4. Vijai Shanker, "Gandhi on Environment", Gandhi Marg, Gandhi Peace Foundation, New Delhi, January-March, 1997.

Gender and Environment

19

Environment, Gender and Development

Prof. Chandrakala Padia*

Today humanity faces a crisis of global dimensions. It is a crisis that not only threatens the survival of the poorest on billion inhabitants of this planes, but a crisis that threatens all of humanity and which has come to be known as the global ecological crisis. From the degradation is proceeding at an unprecedented rate—all in the name of development. Development or economic progress has always produced, as a side effect, the degradation of the environment. The more the material progress in general, the more deterioration, This has been an integral part of the developmental process so far, because we have hitherto taken a compartmentalized view of the situation.

As a consequence of this there is paradigm shift from this compartmentalized view to a holistic perspective; and from emphasis on 'Development' to 'Sustainable Development'. Now people realize that Environment and Development are not and need not be incompatible. They are complementary to one another. Protection of Environment is an important aspect of economic, political and social development. Environment and Development being two sides of the same coin, a pragmatic strategy dealing with environmental challenges has to be an integral part of the development process. The attitude of some economists treating

* Professor, Department of Political Science, Banaras Hindu University and Co-ordinator Centre For Women's Studies & Development, Banaras Hindu University. Varanasi.

Environment as an externality to the production process, which is difficult to quantify in terms of normal economic accounting is no longer justified and in future, environment impact of development projects has to be taken as seriously as production targets. In fact human development depends on the carrying capacity of the ecological systems. It can take place only on the foundation of the continued maintenance of the stability, adaptability and resilience of the dynamic ecological systems. For, only the humans—the most intelligent of all living creatures of Nature—have the moral and legal responsibility to effectively play the role of the saviour, guardian and protector of the environmental (natural), economic, cultural and political sustainability at all levels—local, national and international and global.

In fact there are four approaches towards 'Nature' The first is that of 'resource conservationism', which treats Nature as essentially a provider of raw materials, and therefore to be husbanded for future economic growth (Khan, 1978; Kahn and Simon, 1984; Stroup and Baden, 1983; World Resources Institute 1986). It is, of course, entirely' anthropocentric': Nature is 'useful-to-Man'. It often assumes continuing economic growth and 'progress', with no major shifts in institutions or policies (except perhaps on population issues in underdeveloped countries). And it often has overtones of technocratic. 'environmental management' from above, or indeed from outside'. 'Resources' are to be defined and conserved by elite expert agencies, which will thwart the efforts or ignorant locals to squander them.

The second major grouping on our spectrum is that of the 'environmental protectionists' (Brundtland, 1987; Scientific American, 1990). Of respectable antiquity, and now dominant in the environmental establishment, they too are 'anthropocentric', but their concerns are far wider than those of the 'resource conservationists'. Initially they are worried over public health. This led inevitably to a focus on atmospheric and water pollution, but also led to an interest in more general environmental issues, for example the need for public parks ('areas of outstanding natural beauty') and for public education (e.g. on diet).

Third we move on to greener, and more radical, territory. Here, among the 'mainstream Greens', we find more varied and quarrelsome groups, and also more original thinking. The Appropriate Technologists, for example, busy themselves not only with finding small-scale technofixes, but with new models of what economic development might mean in backward rural areas (Carr, 1985; McRobie, 1981). Among the 'mainstream Greens' the strees, in general, is on 'sustainability' (Brown, 1981; Daly, 1973: Schumacher, 1973). Sustainability, not pollution in itself, is the essential limits to growth. In fact, 'Green' economists reject the whole traditional concept of growth, preferring a 'steady-state' economy which is as self-sufficient as possible. Hence there is stress on renewable energy, on recycling of waste into raw materials, and on meeting 'real human needs' rather than Gross National Product growth targets. Both market capitalism and State socialism are deplored, and seen as essentially exploitative and oppressive.

Finally, we come to the furthest reaches of deeper Green persuasion, where 'ecocentrism' dominates, where wilderness is prized for its own sake, where 'ecosystem collapse' is seen as imminent, and where the Higher Pagans celebrate the ancient cult of Gaia (Devall and Sessions, 1985; Naess, 1973). For these groups, the industrial system is destroying the planet. It must be stopped, or the earth will wreak revenge. 'Growth' means cancer. nothing more; a cancer that threatens to spread worldwide, and destroy all life. No attempt should be made to 'develop' the Third World; people there are fortunate to have escaped industrial civilization. Some advocate a return to hunter gatherer society and a collapse in human population. All regard Nature as the found of stability, interdependence and co-operation (rather than of conflict, as the Social Darwinists had held). The 'spiritual' dimension is strongly emphasized: Nature is not to be preserved as much as worshipped, and all life should be led in harmony with her laws. 'Anthropocentrism', even in its benign form as 'stewardship', is derided: humans are not, and never have been, in charge of the earth.

In this entire discourse, there is very little attention paid to Women's relationship with the Environment. Is it distinct from that of men's? This is and areas which needs innovative conceptual work and practical interventions. Most of the work done by the World Bank of environmental issues is totally gender 'neutral', which in practice usually translates into a male bias. There is an implicit assumption that the environmental and environmental issues are perceived in the same way by males and females. Given the context of men's and women's differing experiences with the environment and their often divergent needs from the natural environment, this assumption may be ill-founded. As Bonnie Kettel (1993) argues, both 'gender' and the 'environment' are social and cultural constructions. Perceptions of 'environment' are thus dependent on human interaction with and use of the natural landscape. From this perspective, perceptions of environment are closely linked to the sexual division of labour, though mediated by cross-cutting factors such as class, ethnicity and race.

In a recent paper on sustainable development, CIDA recognized a clear connection between gender equity and environment based both on women's close relationship with natural resource bases (trees, water, land) and on the central role of women in economic production (CIDA 1991). Nonetheless, this recognition is still far from being generally accepted either by scholars writing on environment and sustainability issues or by development agency personnel. For example, Micheal Redclift's otherwise insightful analysis of the relationship between global environmental crisis and capitalist expansion and the linkages between North and South gives only cursory attention to the issue of gender and different attitudes held by men and women towards the environment (1987).

An emerging feminist analysis of the environment has drawn attention to significant linkages between women's roles, the global commoditization of agricultural production and environmental degradation (Dankelman and Davidson 1988; Mackenzie 1993).

Jane Collins (1991) suggests that the literature on women and environment has often treated women as a homogenous group, overlooking differences of class, or ethnicity. Consequently it has tended to generalize about the role of women in resource management and to minimize rigorous analysis of how women's interaction with environment has changed over time and in varying situations. In this sense, the literature on environment has not benefited from the earlier insights of scholars writing on the sexual division of labour and responsibility in agriculture (see also Bina

Agarwal 1911). Similarly, development agencies have tended to have a slow reaction time. Although the significance of male migration (and increasing female migration) has been acknowledged widely for over a decade, few agencies have reformulated their programming to take into account the constraints faced by women heads of households.

The hesitation to link gender and environment in a systematic way reflects a general tendency for development analysts to strive for clear boundaries among and between concepts, in over better to understand aspects of the whole, and only later to attempt to make explanatory linkages among them. It is easier to begin with an analysis of "environment" and later to consider 'gender', than to begin with an analysis of 'gender and environment'. As a general rule, we have tended to organize different experiences into separate analytical categories. This can be a problem. For example, Kettle argues the perceptions of the environment are closely related to the sexual division of labour, and that since much of women's environmental knowledge is associated with their knowledge has also been undervalued. Significantly, the components are not equal to the whole. An analysis of women's agricultural work does not yield a thorough understanding of the nature of women's relationship with the environment and the depth and breadth of their knowledge of environmental issues. Similarly, a focus on women's environmental knowledge per se may overlook or undervalue the complexities of the work tasks themselves and the extent to which knowledge is closely linked to them (Kettel 1993). Elizabeth Ardayfio-Schandorf's work in northern Ghana reveals the close link between women's knowledge of forest resource (1993). However this instrumentalist knowledge is situated within what might be termed women's 'meta-knowledge, adopted by most development specialists, undermines serious consideration of this meta-knowledge, and consequently impoverishes both development thinking and practice.

This close link between women and environment is now termed as ecofeminism. This concept is based on four arguments. (1) There are important connections between the domination and exploitation of nature. (2) In patriarchal thought, women are identified as being closer to culture. Nature is seen as inferior to culture, hence, women are seen as inferior to men. (3) Because the domination of women and the domination of nature have occurred together, women have a particular stake in ending the domination of nature, 'in healing the alienated human and non-human nature'. (4) The feminist movement and the environment movement both stand for egalitarian, non-hierarchical systems. They thus have a good deal in common and need to work together to evolve a common perspective, theory and practice.

In the ecofeminist argument, therefore, the connection between the domination of women and that of nature is basically seen as ideological, as rooted in a system of ideas and representations, values and beliefs, that places women and the non-human world hierarchically below men. And it calls upon women and men to reconceptualize themselves, and their relationships to one another and to the non-human world, in non-hierarchical ways.

Others such as Ynestra King and Carolyn Merchant argue that the nature-culture dichotomy is a false one, a patriarchal ideological construct which is then used to maintain gender hierarchy.

Ecofeminist discourse therefore, highlights (a) some of the important conceptual links between the *symbolic* construction of women and nature and the ways of acting upon them (although Merchant alone goes beyond the level of assertion to trace these links in concrete terms, historically); (b) the underlying commonality between the premises and goals of the women's movement and the environmental movement; and (c) an alternative vision of a more egalitarian and harmonious future society.

Vandana Shiva's work on India takes us a step forward. Like the ecofeminists, she sees violence against nature and against women as built into the very mode of perceiving both. Like Merchant, she argues that violence against nature is intrinsic to the dominant industrial/development model, which she characterizes as a colonial imposition. Associated with the adoption of the developmental model. Shiva argues, it was a radical conceptual shift away from the traditional Indian cosmological view of (animate and inanimate) nature as *Prakriti,* as 'activity and diversity' and as 'an expression of *Shakti,* the feminine and creative principle of the cosmos' which in conjunction with the masculine principle

(*Purusha*) ... creates the world'. In this shift, the living, nurturing relationship between man and nature as earth mother was replaced by the notion of man as separate from and dominating over inter and passive nature. 'Viewed from the perspective of nature, of women embedded in nature', the shift was repressive and violent. 'For women ... the death of *Prakriti* is simultaneously a beginning of their marginalization, devaluation, displacement, and ultimate dispensability. The ecological crisis is, at its root, the death of the feminine principle ...'.

At the same time, Shiva notes that violence against women and against nature are linked not just ideologically but also materially. For instance, Third World women are dependent on nature 'for drawing sustenance for themselves, their families, their societies,' The destruction of nature thus becomes the destruction of women's sources for 'staying alive'. Drawing upon her ecperience of working with women activists in the Chipko movement—the environmental movement for forest protection and regeneration in the Garhwal hills of north-west India—Shiva argues that "Third World women' have both a special dependence on nature and special knowledge of nature. This knowledge has been systematically marginalized under the impact of modern science: 'modern reductionist science, like development, turns out to be a patriarchal project, which has excluded women as experts, and has simultaneously excluded ecology and holistic ways of knowing which understand and respect nature's processes and interconnectedness as *science*'.

Feminist Environmentalism

I would like to suggest here that women's and men's relationship with nature needs to be understood as rooted in their material reality, in their specific forms of interaction with the environment. Hence, in so far as there is a gender and class (caste/race)-based division of labour and distribution of property and power, gender and class (caste/race) structure people's interactions with nature and so structure the effects of environmental change on people and their responses to it. And where knowledge about nature is experiential in its basis, the division of labour, property and power which shape experience. Indeed, on the basis of their experiential understanding and knowledge, they could provide a special perspective on the processes of environmental regeneration, one that needs to inform our view of alternative approaches to development.

Class-Gender Effects

We come then to the class-gender effect of the processes of degradation, statization and privatization of nature's resources, and the erosion of traditional systems of knowledge and resource management. These processes have had particularly adverse effects on poor households because of the noted greater dependency of such household on communal resources. However, focusing on the class significance of communal resources provides only a partial picture—there is also a critical gender dimension, for women and female children are the ones most adversely affected by environmental degradation. The reasons for this are primarily three-fold. First, there is a pre-existing gender division of labour. It is women in poor peasant and tribal households who do nuch of the gathering and fetching from the forests, village commons, rivers and wells. In addition, women of such households are burdened with a significant responsibility for family subsistence and they are often the primary, and in many female headed households the sole, economic providers.

Secondly, there are systematic gender differences in the distribution of subsistence resources (including food and health care) within rural households, as revealed by a range of indicators: anthropometric indices, morbidity and mortality rates, hospital admissions data, and the sex ratio (which is 93 females per 100 males for all-India). These differences, especially in health care, are widespread in India (and indeed in South Asia).

Thirdly, there are significant inequalities in women's and men's access to the most critical productive resource in rural economies, agricultural land and associated production technology. Women also have a systematically disadvantged position in the labour market. They have fewer employment opportunities, less occupational mobility, lower levels of training, and lower payments for the same or similar work.

On the basis of above, it becomes clear that gender perspective must come in any analysis of environment, for it ultimately provides a deeper analysis and makes demands for changes that will effect more profoundly the structure of current social, economic and cultural process. It leads to a fundamental re-examination of social structures and institutions, to a rethinking of hierarchical gender relations and ultimately to the loss of power of entrenched elites, which will effect some women as well as men. It focuses on both the condition of women, i.e. their material state in terms of education, access to credit, technology, health status, legal status, etc. and the *position* of women, i.e. the more intangible factors inherent in the social relations of gender and in the relations of power between men and women.

Protection of Environment is an important aspect of economic, political and social development. Environment and Development being two sides of the same coin, a pragmatic strategy dealing with environmental challenges has to be an integral part of the development process.

In this entire discourses, there is very little attention paid to Women's relationship with the Environment. Is it distinct from that of men's? This is an area which needs innovative conceptual work and practical interventions. Most of the work done by the World Bank of environmental issues is totally gender 'neutral', which in practice usually translates into a male bias. There is an implicit assumption that the environment and environmental issues are perceived in the same way by males and females.

Elizaeth Ardayfio-Schandorf's works in northern Ghana reveals the close link between women's knowledge of forest resource (1993). However this instrumentalist knowledge is situated within what might be termed women's 'meta-knowledge' of forest resource, which includes both natural science as well as cultural and symbolic knowledge. It is this met-knowledge, adopted by most development specialists, undermines serious consideration of this meta-knowledge, and consequently impoverishes both development thinking and practice.

We come then to the class-gender effect of the processes of degradation, statization and privatization of nature's resources, and the erosion of traditional systems of knowledge and resource management. These processes have had particularly adverse effects on poor households because of the noted greater dependency of such households on communal resources. However, focussing on the class significance of communal resources provides only a partial picture—there is also a critical gender dimension, for women and female children are the ones most adversely affected by environmental degradation.

20

Women and Ecological Problems in Uttaranchal

Dr. Girish Chandra Pande*
Dr. Aparna Joshi**

The hill state of Uttaranchal comprises 13 districts of Kumaun and Garhwal, two important administrative and linguistic divisions of the newly created state, stretching between 28°44′ and 31°81′ north and 78°35′ to 81°5′ east is a typical case of how development has led to ecological imbalance. Kumaun division consist of Almora, Pithoragarh, Nainital, Bageshwar, Champavat and Udham Singh Nagar districts whereas Garhwal division consists of Uttarkashi, Chamoli, Pauri, Tehri, Radarprayag, Dehradun and Hardwar districts. Except the Terai plains of Udham Singh Nagar district and some areas of Dehradun and Hardwar districts, the newly created state of Uttaranchal lies within the Himalayan mountain system. Geo-tectonically, the region has a very fragile eco-structure. The twin process of forest destruction and wide spread use of explosives for road extension and mining have accelerated the rate of soil erosion and land slide creating severe ecological imbalance.

The current ecological crisis in the region has a long history. For more than a century, various developmental activities going on

* Senior Lecturer, Department of Sociology, K.G.K.P.G. College, Moradabad.

** Lecturer, Department of Sociology, G.D.H.G.P.G. College, Moradabad.

in the region have brought about major ecological transformations in this region, which are affecting the lives of women of the region. Some of the major problems plaguing this region are heavy male migration to the plains and money order economy to bridge the subsistence gap, excessive commercial exploitation of the forest for defence and developmental needs, construction of river valley projects, industries and communication network creating an acute scarcity of fuel wood and fodder, over felling of trees causing floods, soil erosion and land slides affecting agricultural productivity.

The worst victim of the ecological crisis in the state are rural women of peasant household who bank upon forests and natural resources for their survival. They are dependent on natural resources like fuel wood, fodder, water, biomass etc. for livelihood. Here observation of Vandana Shiva, a noted environmentalist appears apt in this regard that the violence to nature, which seems intrinsic to the dominant development model is also associated with violence to women who depend on nature for drawing sustenance for themselves, there families, their societies, (Shiva 1988) The ecological destruction has affected rural women of the area in a catastrophic way by increasing their work loads and consequently reducing their status and affecting their health. Heavey male migration from the region has further worsend the condition of already overburdened women of the region. "The special difficulties that women face-their extraordinary work burden or lack of access to health care, for instance of not arise out of ecological deterioration per se. They are located in the sexual division of labour, marked by a double work burden (at home and outside) and by the specific nature of tasks they do, and in the unequal distribution of resources like food within the household, which stem from women's inferior status in the household and from a lack of control over cash and productive resources like land. But given this situation, environmental destruction exacerbates women's already acute problems in a way very different from men's." (C.S.E. 1985)

The women of the state are overburdened with subsistence agriculture and cattle care in addition to the drudgery of bringing fuel wool, fodder and water from long distances and other domestic responsibilities. More than 95 per cent women are cultivator in the Hill Sate of Uttranchal. The uneconomic and fragmented land holdings are managed by women with a little help from men-folk. Culturally accepted division of labour within the family entrusts upon them, collection of fuel wood, fodder and water fetching. The reduced availability of biomass and other natural products fuel to ecological degradations compel them to cover long distances for collecting fuel wood, fodder and water per women. Detailed analysis of different activities performed by the rural women in the hill economy and time devoted to them has been exhibited in the Table 20.1.

Table 20.1

Activities	*Male/ Female*	*Time spent by per women Hr./day*
Agriculture		4.11 Hrs.
Field Cleaning	F	
Ploughing and Horrowing	M	
Putting Manure	F	
Levelling and Weeding	F	
Transplanting and Harvesting	F	
Threshing and Winnoing	F	
Transporting	F	
Animal Husbandry	F	.73 Hr.
Water Fetching	F	1.42 Hrs.
Fodder and Fuel Wood Collection	F	4-5.41 Hrs.
Domestic Drudgery	F	4-5 Hrs.

Sources: Various studies and field work.

Inspite of the fact that the agriculture is not even subsistence in the existing situation, even then it is largely predominant in Uttaranchal. But it is by no means easy and simple. It requires very diligent and intensive labour "Every male cultivator here needs a wife to perform the drudgery of the field". (Pant 1935) Most of the agricultural tasks except ploughing and harrowing (taboo for women according to Dharmshashtras) are performed by the women of the state. Due to heavy male migration most of the household are left with no able bodied men to help her in agriculture. Therefore most of the agricultural work right from preparing field to threshing and winnowing are the responsibilities of women. "The few men left behind are not really of much help as they themselves admit without much shame that their work normally finishes after ploughing the fields. (Khandekar 1986).

Animal husbandry and Fodder collection are also duties of women in rural household. The women are required to look after cattle, clean cattle sheds (goth) and take them to grazing. Paucity of fodder requires women to take long march to collect it. Forest department is also contributing to make the problem more acute for women by banning grazing and collecting fodder in the forest.

Water fetching is one of the most strenuous activities in the hill areas which is assigned to women and it consumes lot of time and energy. Different studies give the evidences that women spend long hours in hill region of Uttranchal in fetching water. Women have to fetch water for cooking washing, cleaning, and drinking for cattle etc. from as much as 2-3 Km. away climbing up and down an average 200-400 ft. everyday. The availability of water is directly related to the level of biomass available in the surrounding environment. In the hill region rich humous of the Oak Forest helps in storing water. The destruction of oak forest and changes from oak to chirpine plantation is responsible for perennial springs to seasonal one resulting long march in search of water for women. Fuel wood collection is also exclusive responsibility of women folk. A study conducted by Swaminathan Madhura in Garhwal hills shows that one or more members of each household daily walk 5-km. steep uphill to collect fuelwood. Collection of fuel for hearth and fodder for cattle alone take nearly 4 to 5 hours. (Jain 1984).

The dependence of women folk in rural area of Uttaranchal for livelihood on forest brought women much closer to forest and as a result these rural peasant women took part in the Chipko movement of the region. Women who produce most or all of subsistence goods wish to maintain the status quo by retaining the traditional eco-system. Women see the question of forest conservation as one of their survival. (Jain 1984) Thus it is implied that women's consciousness of the environment and therefore their conservative approach is then not something inherent, arising out of their being women as is often propagated. It is rooted in their roles and work tasks, which lead them to interact extensively with natural resources for survival (Venkateswaran 1995).

The problem of women of Uttaranchal region does not end with her fight against exploitation of forest and natural resources but the great problem that the migration of male has brought ironically is borne by the women left behind. The penetration of cash economy has created a new culture, which has psychologically alienated men from their eco-system. The other reasons assigned to heavy male migration from the region are primarily prevailing structural conditions at the origin which operates as push factors or a combination of both push and pull factors in which push factors dominate. (Khanka 1988)

Social and psychological impact of migration can't be quantified. Women of migrant household face many tensions, pressures and anxieties. The worst sufferers of migration are young married women who suffer from loneliness and long separation. Export of able-bodied men from this region means more work for women, making women's plight more miserable. The drudgery and migration of men-folk in search of work, leaving women alone to cope with strenuous life of the village where they lead their life as quasi-widows is not a new phenomenon for them. (Weber 1987)

Finally the extraordinary workload also affects the health status of women. They eat less food that is required consequently majority of them are malnourished. Besides, the years of hard work, over exertion and malnutrition make her vulnerable to a number of diseases, which cause premature death. The health facilities are very scarce in rural areas of Uttaranchal and even those little facilities are available are not availed by women because they don't have enough time to seek health care. A study conducted by Pande (1999) shows that hill women work for 16.49 hours a day. The worst condition of women's reproductive health can be gauged by the fact that childbirth takes place at homes, are assisted by the untrained elderly women. They continue their back breaking work almost until childbirth and often immediately after childbirth.

In the final analysis we may submit that the pitiable predicament of rural women in Uttarnchal is owing to unfair division of labour based on gender and ecological degradation. The problem of women in the region has multiplied manifold due to heavy migration of able-bodied men, which leaves women alone to cope with harsh conditions of life prevalent in the region.

REFERENCES

1. C.S.E., (1985), The State of India's Environment 1984-85, The Second Citizen's CSE Report, New Delhi.
2. Jain Sobhita 1984, Women and People's Ecological Mevement. E & PW Vol. 19, A Case Study of Women's Role in the Chipko (41) Oct. 13, Movement. in U.P., 1984.
3. Khandekar Sreekant, Kumaun & Garhwal, The Male Migration *India Today* 1986, Dec. 15, 1986.
4. Khanka S.S., 1988, Labour Force, Employment and Unemployment in a Backward Economy, Delhi HPH .
5. Pande P. N., (1999), Drudgery of the Hill Women New Delhi Indus Pub.
6. Pant S. D., (1935), The Social Economy of the Himalayas, London.
7. Shiva Vandana, (1988), Staying Alive: Women, Ecology and Survival in India, New Delhi, India, Kali for Women.
8. Venketswaran S., (1995), Environment, Development and the Gender Gap ND Sage.
9. Weber Thomas, (1987), Hugging the Trees: The Story of Chipko Movement N.D. Viking.

Economic, Management and Scientific Perspectives

21

Environmental Economics

Challenges and Suggestions

Dr. Anup Kumar Mishra*
Prof. G.S. Kushwaha**

Introduction

Environmental degradation or pollution has always been a cause of concern universally, but only in recent times the need for a clean environment has necessitated an indepth probe into pollution and its allied problems. The transformation of clean environmental quality from free to a scarce good is, paradoxically, attributed to man's success in achieving unprecedented economic plenty. The high levels of pollutants that strain the earth's natural absorptive capacity today are the other side of the coin of the high levels of output growth in all countries.

According to the world development report 1992 developing country output would rise by 4-5 per cent a year between 1990 and 2030 and by the end of the period it would be about five times what it was in 1992. Industrial country output would rise more slowly but would still triple over the period. World output by 2030 would be 3.5 times of it was in 1992 (a roughly $69 trillion in 1990 prices.)

* Lecturer in Economics, D.A.V. Degree College, Varanasi.

** Retd. Head & Dean, Faculty of Social Science, Banaras Hindu Unitersity, Varanasi.

If environmental pollution and degradation were to rise in step with such arise in output, the results would be appealing environmental pollution and damage. Tens of million crore people would become sick or die each year from environmental causes. Water shortage would be intolerable and tropical forests and other natural habitats would decline to a fraction of their current size. Hence the protection of the environment is an essential ingredient of development, without which development cannot qualify to be called 'development' and therefore there is impending need to integrate environmental considerations in the development policy making.

The environmental issues that countries face vary with their stage of development, the structure of economics and their policies on ecological aspects. In general some environmental problems are associated with their lack of economic development while many are associated by the growth of economic activity.

Theoretical Aspects of Environmental Economics

Adam Smith and his followers in the classical school were concerned mostly with resource economics and not environmental economics. The reason may be that resource problems were more presenting than environmental problems during the 18th and 19th centuries; Neo-classical economists, who developed the modern theory of the firm, the consumer and market equilibrium in economics, like Alfred Marshall, Knut Wicksell, L. Walrus and A.C.Pigou developed the concepts of external economics and diseconomies which are crucial in understanding environmental problems (Michael Common, 1998, Environmental and Resource economics, Longman, New York).

Environmental economics mainly builds its theoretical foundation on welfare economics. Environmental economics are considered as problems of non-optimality. Because, environmental services and functions treated as free goods results in 'market failure' leading to non-optimality. The ideas of optimality and non-optimality are the core of the subject matter of welfare economics. Welfare economics attempts to asses what an optimal configuration of the economy is, what factors cause deviations from such optimality and how to correct such deviations. Environmental pollution causes such deviations from optimality (Meade, 1973, The Theory of Economic Externalities, Institute University de Hautes Etudes International, Geneva).

Hence, it is necessary to mention about Pareto optimality that is a necessary condition for attaining maximum social welfare. In the Pareto optimum position the welfare of an individual of the society cannot be increased without decreasing the welfare of another. The position of this Pareto optimum cannot be achieved even under perfect competition, if pollution exists there. This is mainly because, when pollution exists in production, private marginal cost will be lower than the social marginal cost, since the former will not take into account costs or harms imposed on others. Therefore when pollution present, equating price with marginal cost will result in over production of the product, that is, more than socially optimum output will be produced (Meade, 1973).

So, due to the non-inclusion of external cost in the calculation of private cost, private market system fails to allocate resources efficiently. Pollution that causes such distortions, hence require special considerations. In economics, pollution is a kind of negative externality or deter mental externality and therefore such externality creates cost burden to firms affected people, pollution controlling authority government and others.

Developing Countries and Pollution

It is an illusion to think that environmental problems affect only rich countries. On the contrary, they are more severe and more ominous in poor countries. The air of Delhi or Kolkata in India are not less dirty than any rich country's city. All the major rivers including river of religious faith, 'Ganga', has also been polluted in India. India has stronger reason to concern about environmental quality. Various kinds of the environmental deterioration have negative bearing on the quality of life, especially on the poor. 1st and 2nd citizens report on the state of the India's environment published by the center for Science and environment (New Delhi, 1982 & 1988) proves the nature of environmental deteriorate in the country and confirms the fact that poor suffer the most. Their argument is, poor may or may not bear short-run cost of environmental protection, but it is no doubt, they have to certainly bear more of the long-run costs of not protecting the environment.

Major Challenges

In May 1984, the world Resources Institute staged an international meeting at Wye, Maryland, to take a fresh look at global resource and population problems and opportunities. This meeting was called as the 'Global Possible Conference', and the conference threatened about the environmental problems in the world and strongly supported the need for sustainable development. At that time, Global Possible conference suggested sustainable future is attainable.

Global possible conference suggested that population can be stablized before it doubles again. Agricultural production can expand to meet all future demands. Economic growth can be sustained with markedly lower energy input. Environment pollution can be markedly reduced as the quality of the environment preserved by more efficient resource recovery and source abatment.

But the Global Possible conference could not spread its message, a faster, as did Rio Conference. In June 1992, the United Nations Conference on Environment and Development (UNCED). popularly known as the Earth Summit took place in Rio. The Earth summit marks a historic turning point in the long struggle to increase international awareness of the true nature of the global environmental crisis. More than 170 leaders of countries participated in the conference and committed their countries to environmetally sustainable development. The Rio Declaration stated :

"... environmental protection shall constitute an integral part of the development process and cannot be considered apart from it.'

Also emerging from the Rio Conference was Agenda 21, a Master Plan of action for moving towards sustainable development. The action plan includes sections on poverty, changing consumption patterns, international trade and cooperation, promoting sustainable agriculture, rural development, desertification and deterestation, protection of atmospheres and the role of NGOs. The Rio Conference is shifted the debate 'Growth Versus Environment' to Growth and Environment' which is the real goal for sustainable development.

Agenda 21—A Positive Step

The preamble of Agenda 21 stages that humanity is up against poverty, hunger, ill-health, illiteracy and the continuing deterioration of the ecosystems. Integration of environment and development concerns would lead to the fulfilment of basic needs, improved living standards for all, better protected and managed ecosystems and a offer of more prosperous future. Agenda 21 reflects a global consensus and political commitment at the highest level on development and environment cooperation.

The most important part of Agenda 21 is its Section I, which covers its social and economic dimensions. The conference called for an international climate which would promote sustainable development through trade liberalization, making trade and environment mutually supportive, providing adequate financial resources to developing countries and dealing with international debt and encouraging macro-economic policies conducive to environment and development.

Current Problems and Role of Economists

Each year the world adds the population equivalent of a new Mexico, mostly in the relating poorer countries. These increases in population have strained the abilities of economies in the developing countries to provide employment, to increase agricultural production and make the needed investments if expanding cities are to carry on properly.

In fact the strain on the earth's atmosphere is excessive and if a further billion people are to lead what is seen as ideal lifestyle, the earth would not be able to withstand this enhanced burden. We know that unchecked growth in consumption is not possible, but there must surely be a better way to stem this growth then through an economic slump.

How is the rapid growth in the world's population to be checked, is a big question. The world's rich can help the poor, but only upto a point. Then what to do?

Economists, more than anyone else belong to the first group with its conservatively optimistic view of the future. We should not see economy or growth as an enemy of environment rather it should be friendly. But, as economists, we do not argue that environmental problems cannot occur. To do so would be to ignore the fact that this do occur. We do argue that environmental problems are not the consequence of economic growth as such, but that they are the consequence of inappropriate patterns of economic activity which would not arise, at any level of economic activity, if economy environment relationships were determined by a properly functioning price mechanism.

The task of economists with respect to natural resources and environmental policy is to provide a framework within which the duty provided by experts in other fields can be viewed and the social implications drawn. Economist can only repeat, without quite understanding, what geologists, ecologists, public health experts and other say about physical and physiological facts. Their craft is to perceive how economics and people in general will respond to those facts.

Economic factors do not block the attainment of a more desirable and sustainable future—they favour it. The payoffs due to better management of resources are large. The present cost of switching to better policies and programmes are small relative to future costs or the damages, that will result if action is not taken.

The industrial world is now realising that the costs of properly disposing of hazardous wastes or recycling them is small relative to the clean up costs of inadequately regulated hazardous waste sites. So the elements of a programme to promote sustainable development support each other and are attractive both environmentally and economically.

Present Situation

Nine years after the Earth summit and the plan of action as embodied in Agenda 21, the general feeling both in developed and developing countries is that things are not moving towards the vision as articulated at the Earth summit. In the earlier UN conference in 1972 and in Global possible conference in 1984 also nothing much happened. Dr. Mustafa Tolba, in his state of the Environment Report 1972-92, said that "time was running out" and that "none of the environmental problems identified by the

Stockholm conference twenty years ago have really have been solved and some of the key issue addressed in 1972 have gone from bad to worse".

The main problem is that developing countries in the process of industrialisation and economic growth are also being afflicted by the same kind of problems as developed nations have. They are eager for economic growth, employment and meagre wages are imparting toxic industries and toxic waste in their countries. But the basic question is that whether the new model for development is acceptable? We must be very careful while adopting this model. The question was very alarming and right when Gandhiji has asked "why must Indian become industrial in the western sense? The western civilisation is urban. Small countries like England and Italy may afford to urbanise their systems. A big country like America, but with sparse population, cannot do otherwise. But one would think that a big country with a teeming population, with an ancient rural tradition, which has hitherto answered its purposes, need not-must not copy the western model. What is good for one nation situated in one condition, is not necessarily good for another differently situated. One man's food is another man's poison."

This is very encouraging factor that United States now accept his responsibility and it is indeed creditable that the fromer Vice-President, Al Gore blames America and the developed world in the right measures for creating the environmental problems. When Al Gore quote our Mahatama Gandhi's words "we must be the change we wish to see in the world" (Al Gore in earth in Balance, 1993, USA) then that was the right time to act upon sustainable development. But we ignored, and if we continue to ignore the concept of sustainable development, the hidden dangers like poverty, ill health, spread of diseases and the latest form of terrorism (11th Sept. 2001 at world trade tower, USA) will take place. So let us not act like OSTRICH. Let us start thinking of an equal world through curtailing the high mass consumption and diverting our resources to poors.

Suggestions for Sustainable Development

To accelerate sustainable development the developed and the developing nations will have to take some bold steps. Some of the fruitful suggestions are:

1. Poverty should be declared as the biggest pollution of the modern society;
2. The major gap of income between developed and developing nations should be filled up;
3. Industrialised countries should reduce quantitative restrictions on labour-intensive manufactures from the third world;
4. Capital flows in the form of development assistance to low-income countries should be enhanced by developed nations;
5. Developed countries should introduce the new products and technologies that foster economic progress without environmental degradation;
6. The developing nations should stablize their population;
7. Over consumption should be stopped for preserving natural resources, for future generations;
8. Depreciation on natural resources should be provided for in the GNP of every nation;
9. Mass literacy and environmental education should go hand in hand;
10. All the funds for the world environment protection should be raised by the developed nations;
11. The developed sectors of the developing countries should take responsibility of environmental education in the country; and
12. Last but not the least is that, men, women and children must be the centre of attention with development woven around people, not people around development.

Conclusive Remarks

Both the developed and developing nations, are responsible for environmental pollution. Infact the economic pattern of the world is not favourable to the balanced growth and is not environment friendly.

It would be unfair to say that the entire responsibility for environmental disaster is that of southern countries. It is not just poverty and underdevelopment which lead to destruction of environment, more to blame are those rich consumer societies who have created the demand for new products. Developing nations should also not only blame rich countries for envrionmental degradation and inequalities of income. They should stop increasing population. After all, the world's rich can help the poor, but only up to a point. Our responsibilities must be our's.

The industrialised nations are also duty bound with regard to other problems that result from excessive population growth. In a sea of poverty, underdevelopment and environmental destruction, there can be no island of affluence. as is the case with all global dangers we face, is not a question of whether it concerns the 'First' or the 'Second' or the 'Third world', it is a question that concerning our 'one world'. We Indians are the first who always talk about the sustainable development from the ancient times. The '*Shloka*' of our holy epic *'Gita* gives the intensive message for sustainable development.... .

...''महाशनी महापाप्मा

विद्धयेनमिह बौरिणम्।

गीता 3-37

... desire ... an enemy. (Gita 3.37) holds that human desires should be cut down to the size or be so reduced to the minimum that one may prevail upon himself to exchange joy and sorrow for a state of constant tasteless indifference to this world and to its economic expansion. This is the only way for sustainable development.

REFERENCES

1. Al Gore, Earth in Balance, Published by Penguin Group, USA, 1993.
2. Buch, M.N., Environmental Consciousness and Urban Planning, Orient Longman, 1993.
3. Caldwell, Lynton Keith, Between Two Worlds, Cambridge University Press, 1990.
4. Common, Michael, Environmental and Resource Economics, Longman, New York, 1988.

5. Chandra J. N. One Man One World, Bliss and Light Publications, Allahabad, 1978.
6. Chodary, Kamla, Gandhi and Agenda 21, Mainstream, Novermber 26, Annual 1994.
7. Deutschland, Germany, No. 2, April, 1994.
8. E P W, August 13, 1994.
9. Goldsmith, E. and Hildyard, H., Green Britain or Industrial Wasteland, Polity Press, 1986.
10. Hindustan Times, September, 8, 1992.
11. Human Development Report, UNDP, 1990, 1991, 1992, 1993, 1997, 1998.
12. India Today, June, 15, 1994.
13. Jazary Indries, The State of World Rural Poverty, FAO International Fund for Agricultural Development.
14. Mainstream, November 26, 1994.
15. Nand, B. R. Mahatma Gandhi, Oxford University Press, Delhi, 1989.
16. Report on the World Social Situation 1993 (U.N.)
17. Sustaining the Earth: Role of Multilateral Development Institutions, Oversees Development Council, Washington D. C., 1993.
18. Lester R. Brown, State of the World 1994, A World Watch Institute Report on Progress Toward a Sustainable Society, W.W.Norton, New York, 1994.
19. N.ss L. Gupta and R. K. Gurjar, Sustainable Development, Vol. 1 and 2, Rawat Publication, 1993.
20. World Development Report, 1994, 2000/2001.
21. World Development Indicators, 2000, UN Publications.

22

Managing Mismanagement of Environment
A Planned Approach

Mr. Shardul Chaubey*

It has been stated in one of the Plan paper 'India is fortunate in the richness of its natural resources...the abundance and diversity of its living resources. Adequately managed, these and other resources can meet high levels of material needs, now and for all times to come. The degree to which a nation can prosper depends on its productivity, which is the efficiency with which it is able to utilise the resources of the environment to satisfy human needs and expectations. If the gains in productivity are to be sustained, resources must also continue to be available overtime. This requires that, while providing for current needs, the resources base be managed so as to enable sustainable development'. It means there must be sustainable development in harmony with the environment. Towards this end, it would have to be ensured that all development programmes, in all sectors, will take environmental considerations fully into account.

The problems encountered in the field of environment in India arise due to conditions of poverty and underdevelopment as also the negative effects of development programmes which have been badly planned or badly implemented. The whole planning process is aimed at development and the removal of poverty. The need to

* Senior Lecturer, Department of Media Studies, School of Management Sciences, Varanasi.

improve the conditions of our people is pressing; under this pressure many concerned with developmental activities lose sight of environmental and ecological imperatives. Realisation concerning these aspects has been with us for only a relatively short period of time, about a decade and half. The demage being done to the environment, because of the large size of the population and its increase, and scale of developmental activities, is of such magnitude that urgent remedial measures are called for. Official and voluntary agencies must work together to create the needed awareness; indeed, environment is all- pervasive, and the success of our efforts in this area will ultimately call for the involvement of the entire population at all levels. This is a philosophy which must permeate the entire effort in the field of environment.

Environmental management, a term encompassing environmental planning, protection, monitoring, assessment, research, education, conservation and sustainable use of resources, is now accepted as a major guiding factor for national development in India. From the early seventies, India has played a significant role at international forums in delineating and articulating the relevance of environmental concerns in the context of economic development. There has been, over the last decade, a progressive strengthening of official involvement in environmental management in India, with increased scientific, technical, administrative and legislative backup at the Central and State levels.

With the realisation that poverty and the state of underdevelopment led to many of the environmental problems that confronted the nation, came the understanding that it was more rapid development which was the best approach. This development has to benefit people (and particularly the poor) by providing for their basic human needs and rising aspirations. Thus, many of the developmental programmes, and particularly those included in the 20- Point Programme, could indeed be termed as environmental management programmes.

However, another class of environmental problems have arisen as unintended side-effects of the very attempts at development. These had to do with the mismanagement of natural resources, large-scale deforestation, the unplanned discharge of residues and wastes, the handling of toxic chemicals, indiscriminate construction and expansion of settlement activities, etc. It is to this class of problems that the tools and methodologies of environmental planning are primarily addressed.

Review of Progress

It is now being increasingly recognised that environmental factors and ecological imperatives must be built in to the total planning process if the long-term goal of making development sustainable is to be achieved. To provide greater systematic impetus and focus to environmental issues at the Central and State level, new organisational structures have been created. The Government of India set up a Department of Environment in the Sixth Plan. The State/UT Government were also asked to set up structures which could act as focal points for enviornmental considerations in the State Plans; the Governments of Andhra Pradesh, Karnataka, Madhya Pradesh, Punjab, Orissa, Rajasthan, Sikkim, Tamil Nadu, Uttar Pradesh and West Bengal have set up appropriate structures. A number of Ministries/Departments of the Government of India now go into environmental considerations in some detail in their major developmental programmes.

Major activities in the area of environment on which work was initiated or stepped up during the Sixth Plan included: water and air pollution monitoring and control; environmental impact assessment; natural living resource conservation; special projects on wildlife; ecological studies by the Botanical and Zoological Surveys of India; eco-development programmes; environmental research promotion; and environmental information, education, training and awareness.

A country-wide rapid inventory on pollution from large and medium industries has been prepared. A programme on Control of Pollution at Source has been initiated. Minimal National Standards for polluting discharges from specific industries were formulated and control measures implemented in a progressively stringent manner. About 30 per cent of large and medium industries of the country have installed pollution control equipment. A network of about 120 monitoring stations to check water pollution has been created. Zoning and classification of all the 14 major inter-State rivers have been completed to provide a basis for water quality management. A river basin wise inventory for Yamuna and Ganga has been prepared to assess pollution load.

'Project Tiger', a Centrally sponsored scheme was an outstanding success in terms of management; it evolved from a species protection programme to one envisaging protection of the total habitat. The establishment and development of the Wild Life Institute of India and strengthening of programmes within the National Zoological Park (which included breeding of endangered species of wild mammals such as lion-tailed macaque, brow-antlered deer, etc.) received great impetus. A 12-point strategy was adopted by India for Wild Life protection and development in October, 1983, to provide the basis for future plans of Wild Life management.

The Forest Survey of India is preparing a national vegetation map using remote sensing and ground survey methods.

Preparatory work has been done for setting up Biosphere Reserves in a few carefully selected and identified areas which have enormous pristine, genetic diversities, for example, Nilgiri, Namdapha, Nanda Devi and Uttarakhand.

A major traditional weakness of forestry, wild life and other sectors in the area of environment is the poor S & T inputs. To rectify this, the work of the premier survey organisations, namely, the Botanical and Zoological Surveys of India, has been oriented to take into account holistic ecosystem management imperatives. Environmental monitoring centres at Calcutta and Madras have been established to study the impact on living resoruces of developmental activities such as hydroelectric and irrigation projects. The plant and animal resouces have been surveyed at important sites such as those to be designated as biosphere reserves or hitherto unexplored or under- explored areas.

Some major field action projects on Eco-development were farmed out to voluntary organisations in the Shivalik Foothills (Punjab), Joshimath and Dasoli areas of Chamoli District of Uttaranchal, Haldighati in Rajasthan and Auroville in Pondicherry. Eco-development Task Forces consisting of ex-servicemen were deployed for activities such as revegetation of degraded areas, soil conservation work, eradication of weeds, etc. The Eco-development programme is designed to generate public participations in the solution of environmental problems and broadening general environmental awareness, particularly through the involvement of students and voluntary agencies.

In order to promote environmental research, nearly 400 research projects have been sanctioned to universities, R & D institutions and non-governmental agencies. Coordinated, multi-institutional projects in priority areas of heavy metals, microbial degradation of industrial wastes and ethnobiology have been take up; as also a multi-agency post-audit environmental monitoring of the Idukki multi-purpose river valley project. Research relevant to the integrated development of the Western Ghats, Himalayan region and the Ganga basin has also been initiated. One centre of Excellence has been set up at IISc, Bangalore which has, as its primary focus, research on problems of the Western Ghats.

A computerised Environmental Information System (ENVIS) with a network of distributed information centres all over the country, has been started; information relevant to pollution control, toxic chemicals, mining, forestry, flora and fauna are the important subjects covered under this. A variety of 'information products' have been prepared including a directory of nongovernmental organisations in the field of environment.

A broad range of programmes on environmental education, training and awareness were launched; workshops were organised in different parts of the country and nation-wide celebrations were organised on World Environment Day and during Wild Life Week. The first National Environmental Congress and the first National Conference of Legislators on Environment were also held as part of the awareness building programme. The National Museum of Natural History has been playing an important role in imparting environmental education to a wide spectrum of society with particular emphasis on childern.

Evnironmental Planning

Environmental considerations in the planning process and in the implementation of national development strategies must be based on an understanding of the following issues and responsibilities:

- The environment with its component of living and non-living resources, represents the most fundamental building blocks for national development and social well being.
- The environment today is under severe threat from the pressure generated by growth of human and animal populations, poverty and the misuse/ unplanned use of natural resources.
- Stabilisation of human and livestock population is as crucial for environmental management as it is for achieving other plan objectives.
- While many of the country's environmental ills could be corrected by rapid economic growth with social justice, utmost care must be exercised to ensure that development activities which bring about such changes are designed so as not to lead to adverse environmental effects. The nation's planning for economic growth and social well-being in each sector must always take note of the need to protect environmental resources, and where possible, must work to secure improvement in environmental quality.
- Therefore, the primary responsibility for environmental protection must rest with each sectoral authority (ministry, department, development agency, corporate body, municipal council, village panchayat, etc.) which would have to develop formal mechanisms to take account of environmental concerns in policies, plans, programmes, projects and legislations that come under their purview.
- Environmental considerations should form an important element in the criteria for setting developmental targets and assessing plan performance in all sectors. Environmental management must be integral to all development activities.
- The Department of Environment at the Centre, and its counterparts at the State level would essentially serve as catalysts to promote environmentally sound national

development, through provision of management information, technical expertise, monitoring, research and administrative support and, wherever possible, limited financial assistance.

- The prime responsibility with regard to the environment would rest with the various implementing authorities for development programmes and with the community. For this, environmental education and awareness building is crucial.
- Environmental planning/protection/improvement requires a coordinated, highly decentralised approach involving the cooperation and active participation of every segment of society, and most importantly of the political leadership.

Basic Policies Relating to Aspects of Implementation

The experience with environmental management so far emphasises two facts. First, that environmental issues and problems arise in virtually every sector of the economy, and at every level of society.

Second, that sustainable solutions for problems that manifest themselves in one sector may not apply in other sectors. Environmental issues have no administrative, sociocultural or political boundaries. Therefore, the initiatives for tackling environmental issues must emerge from official as well as non-official agencies and individuals, operating at different levels. Success in achieving environmentally sound development will depend greatly on the extent of co-operation that can be achieved between Government (Central, State, Local), its subsidiary agencies, voluntary groups, financial institutions, corporate groups in the public and private sectors, educational and research bodies, professional societies, religious and cultural institutions, etc.

The Eleventh Plan would attempt to remove some of the weaknesses in the existing environmental planning system. Given the close linkages between different subject areas relating to the environment, it is difficult, and often counter-productive, to assign absolute priorities, in the sense of an "order of importance", to these areas.

For a variety of basic economic activities, high priority would have to be given to the management of natural living resources; but these cannot be managed without attention to land and water management. Again, measures to control the growing livestock population are vital for purposeful environmental management but these come under the jurisdiction of authorities different from those who manage other natural living resources (flora and fauna-particularly economic plants, forests, wildlife, fisheries, etc.). Viewed from another perspective, environmental issues relating to human settlements (shelter, potable water supply, waterborne diseases, slums, etc.) deserve very high priority in any agenda of environmental tasks. Therefore, each agency, institution or group must formulate its own priorities for action, among the spectrum of environmental management issues, based upon its own direct responsibilities and capabilities. Environmental authorities such as the Department of Environment will deal with those subject areas for which they have been assigned direct managerial responsibility, e.g., pollution monitoring and control, environmental research and development, etc. The direct goals relating to the subject of environment as a whole would be:

- Institutionalising the process of integrating environmental management and development.
- Inducing organisations at the Central, State and local levels to incorporate environmental safeguards in their plans and programmes.
- Securing greater public participation in environmental management.
- Establishing a strong S&T base for environmental research and development, demonstration and extension activities.
- Strengthening mechanisms for ensuring corrective action with regard to environmental degradation that has already taken place.

The existing framework of institutions which can contribute to the multidisciplinary and multisectoral approaches required for effective environmental management would be fully used. Apart from the large S&T infrastructure under the Central and State Governments in the form of scientific agencies/departments/ national laboratories, etc. and educational institutions, full cooperation will be sought from more than 200 nongovernmental organisations or voluntary agencies working in the field.

Major Programmes

Environmental Programmes were taken up during the Sixth ,Seventh Plan period more in the form of nucleating activities. These would now have to receive a greater impetus in terms of investment and even more through coordinated, expeditious implementation strategies during the Tenth Plan period. Salient programmes to be undertaken, including the thrust areas, are outlined below:

Pollution Monitoring and Control

Environmental pollution is a serious and growing hazard in India. Its impact on human health and well being is both direct, (e.g., inhalation of polluted air and intake of contaminated water), or indirect, by its impact on the health of environmental resources (loss of soil fertility, corrosion of structures, death of aquatic life, etc.).

Waterborne pollution, from both community and industrial sources, is probably the most important health hazard in India. A large part of the population has to depend upon untreated or inadequately treated water supplies.

In the sector of industrial pollution, while the medium and large industries can be induced to enforce pollution control, the problem becomes almost intractable for the vast numbers of small industries that have proliferated unplanned in many parts of the country.

Pollution of the environment from noise, both at the community level and in the industrial work place, is another serious threat to human welfare.

Pollution in the costal zone, resulting in the destruction of valuable living natural and marine resources, and spoiling of tourist attractions like beaches is now attracting growing attention.

The Central Board for the Prevention and Control of Water Pollution spearheads the effort at systematically tackling pollution problems in the country. The basic tasks before the Board are: assessment and control of air pollution; assessment and control of coastal pollution; development of professional expertise and trained manpower; development of cost-effective technologies for air and water pollution control; and strengthening the support for pollution monitoring and control.

Moving Ahead

In the Approach Paper to the Tenth Five Year Plan for the period 2002-2007, the government has recognised that mere economic growth is not enough and that development must be defined in broader terms of human being enhancement (11). The

Tenth Plan has thus outlined a set of monitorable targets to achieve its objectives:

- Reduction of poverty ratio by 5 percentage points by 2007 and by 15 percentage points by 2012;
- Providing gainful high-quality employment to the labour force over the tenth Plan period;
- All children in school by 2003; all children to complete 5 years of schooling by 2007;
- Reduction of gender gaps in literacy and wages rates by at least 50 per cent by 2007;
- Reduction in decadal rate of population growth between 2001 and 2011 to 16.2 per cent;
- Increase in literacy rate to 75 per cent with in the Plan period;
- Reduction of Infant Mortality Rate (IMR) to 45 per 1000 live births by 2007 and to 28 by 2012;
- Increase in forest and tree cover to 25 per cent by 2007 and 33 per cent by 2012;
- All villages to have sustained access to portable drinking water within the Plan period;

- Cleaning of major polluted rivers by 2007 and other notified stretches by 2012 (12).

As indicated, these targets relate to good environmental performance and governance, highlighting how cross-cutting environment management issues are.

Approach to the Tenth Plan

The basic objective of the Tenth Plan is to bring about all round enhancement of human well being by eradicating poverty through adopting well conceived development strategies in which environmental concerns are posited as the vital aspects. environmental management and economic development are mutually supportive aspects of the same agenda. A poor environment undermines development, while inadequate development results in a lack of resources for environmental protection.

Environmental conservation must go hand in hand with economic development because any economic development which destroys the environment will create more poverty, unemployment and disease. Environmentally destructive economic development will impoverish the poor even further and destroy their livelihood resource base. Environmental concern must be linked to people's lives and well being, impact as these do various aspects of human existence tha t include health and livelihood amongst others. Air pollution, soil degradation, deforestation, desertification, shrinking wetlands. Inadequate public health and sanitation, indoor pollution in rural areas, growing water scarcity, falling ground water tables, the lack of minimum flow in the rivers and over extraction of water for irrigation purposes are some of the environmental problems that need to be addressed first before any poverty alleviation programme can meet with success.

Strategy for the Tenth Five Year Plan

The ministry has identified the following strategy to meet the objectives of the Tenth Plan:

(a) To promote 'sustainable development' by emphasising the intrinsic linkage between environment management and socio- economic development of the people, through increased access to natural capital which is an essential component of reducing poverty. The emphasis will be on creating an enabling environment in terms of policies, institutions which support the over-arching objective of poverty reduction.

(b) To mainstream environment issues in the developmental strategy, a multipronged approach will be followed which will include the incorporation of environment in the curriculum, sharpening the mechanism for the environmental auditing of the development activities, creating heightened awareness, broadbasing community involvement and to forging a coordinated effort with corporate and public sector in environmental management.

(c) To input environmental concerns in the developmental scenarios by treating environment as a cross-sectoral issue.

(d) To take concrete steps for increasing forest/tree cover from the existing level of 19.37 per cent to 25 per cent by 2007 and 33 per cent by 2012. While striving for this goal the thrust will be given to:

1. increasing productivity;
2. efficient management of resources; and
3. focus on the local needs and location specific strategies in terms of species selection and management of resources.

(e) Cleaning of major polluted rivers by 2007 and other notified stretches by 2012. In order to have efficient, smooth and time bound attainment of the goal, a holistic and integrated approach of addressing not only the intercepting sewers and sewage treatment plants which currently constituted the core of the Plan, but also other components as internal sewage, solid waste management and toilets with the participation of states, public and local bodies. The focus will on be treating the town as a unit by way of convergence of various schemes.

(f) Major changes in the institutional framework with a view to make it responsive and accountable to perceived needs of people.

(g) Involvment of youth in environmental education, through a National Green Corps scheme to be implemented in all educational institutions covering class 8th to 12th class. (13) bage or any other material (except approved fuel) in the cities and towns to reduce secondary sources of air pollution. It is also proposed to set up a national network of air pollution monitoring stations for the parameters notified in respect of ambient air quality standards and for parameters for which standards have been prescribed for various sources. A database of pollutants releases into the air from different sources will be built up. biomass is a major focus area.

(h) Management of environment-related development problems which include air andwater pollution, and domestic and industrial waste management. The UNDP is also involved in rating different industrial sectors for their environmental performance.

(i) Capacity building at various levels-government, private sector, women's groups, NGOs, all associated with decision-making processes in the area of natural resources and in environmental education.

(j) Sustainable Development Network Programme (SDNP) to make relevant information on sustainable development readily available to decision- makers.

(k) Helping the ministry to meet the control measures as specified in the Montreal Protocol, which has set out a time schedule for freeze and reduction of ozone depleting substances.

(l) UNDP is the lead implementing agency for the Global Environment Facility (GEF) in India. With nearly 15 ongoing projects, India is the second largest recipient of GEF funding. Biodiversity and climate change are the key focal areas for UNDP-GEF with an emphasis mainly on technical assistance and capacity building. Apart from global environmental considerations, GEF projects strongly reflect national priorities, and are implemented in close collaboration with the Government.

REFERENCES

1. Approach Paper To The Tenth Five Year Plan 2002-2007, Planning Commission, Government of India, p. 1.
2. http://www.worldbank.org/pics/pid/in43728.txt
3. Approach Paper To The Tenth Five Year Plan 2002-2007, Planning Commission, Government of India p. 1.
4. Approach Paper To The Tenth Five Year Plan 2002-2007, Planning Commission, Government of India p. 1.
5. *Down to Earth*, Vol. 7, No. 17, January 31, 1999, p. 32.
6. http://www.worldbank.org/pics/pid/in43728.txt.
7. Approach Paper To The Tenth Five Year Plan 2002-2007, Planning Commission, Government of India p. 1
8. Approach Paper To The Tenth Five Year Plan 2002-2007, Planning Commission, Government of India, p. 9.
9. http://www.worldbank.org/pics/pid/in43728.txt.
10. http://www.worldbank.org/pics/pid/in43728.txt.
11. Approach Paper To The Tenth Five Year Plan 2002-2007, Planning Commission, Government of India, p. 3.
12. Source: Approach Paper To The Tenth Five Year Plan 2002-2007, Planning Commission, Government of India, p 3.
13. Tenth Five Year Plan 2002-2007 and Annual Plan 2002-2003, Volume 1, Ministry of Environment and Forests, Government of India, New Delhi, p. 24.
14. Tenth Five Year Plan 2002-2007 and Annual Plan 2002-2003, Volume 1, Ministry of Environment and Forests, Government of India, New Delhi, p. 25.
15. http://www.worldbank.org/pics/pid/in43728.txt.
16. http://www.worldbank.org/pics/pid/in43728.txt.
17. Tenth Five Year Plan 2002-2007 and Annual Plan 2002-2003, Volume 1, Ministry of Environment and Forests, Government of India, New Delhi, p. 26-28.
18. Approach Paper To The Tenth Five Year Plan 2002-2007, Planning Commission, Government of India, p. 31-33.

19. Tenth Five Year Plan 2002-2007 and Annual Plan 2002-2003, Volume 1, Ministry of Environment and Forests, Government of India, New Delhi, p. 30.
20. Tenth Five Year Plan 2002-2007 and Annual Plan 2002-2003, Volume 1, Ministry of Environment and Forests, Government of India, New Delhi, p. 32-33.
21. Tenth Five Year Plan 2002-2007 and Annual Plan 2002-2003, Volume 1, Ministry of Environment and Forests, Government of India, New Delhi, p. 29.
22. http://sdnp.delhi.nic.in/igidr/brochure.html
23. http://sdnp.delhi.nic.in/igidr/brochure.html
24. http://www.undp.org.in/env.htm
25. http://www.ihdp.uni-bonn.de/html/who/natcom/india-publ-sscmeeting.htm
26. http://www.winrockindia.org/
27. www.cseindia.org
28. www.cseindia.org
29. www.devalt.org
30. www. Igidr.ac.in
31. www frlht-india.org
32. TERI Annual Report, 2000/2001, p. 23.
33. TERI Annual Report, 2000/2001, p. 31.
34. TERI Annual Report, 2000/2001, p. 32.
35. TERI Annual Report, 2000/2001, p. 32.
36. TERI Annual Report, 2000/2001, p. 33.
37. TERI Annual Report, 2000/2001, p. 33.
38. TERI Annual Report, 2000/2001, p. 35.
39. TERI Annual Report, 2000/2001, p. 36.
40. TERI Annual Report, 2000/2001, p. 36.
41. TERI Annual Report, 2000/2001, p. 87.

Box: The Small Industries Pollution Problem

1. C M D'Souza, Integrating Environmental Management in Small Industries of India, Electronic Green Journal, Issue 14, Spring 2001.

http://egj.lib.uidaho.edu/egj14/souza1.html

23

WTO and Environment

Dr. Sangeeta Jain*

Almost all nations have recognised the importance of environment and nature, its bio-diversity and the need to maintain it. Boosting the balance of ecosystem has become priority in almost all governmental and non-governmental policy-making organisations and they are making significant efforts in the field of environmental protection and in developing environmental standards even in trade for both products and production processes. Though environmental issues are non-trade issues. Yet they have been taken care of in World Trade Organisation's Agenda.

World Trade Organisation (WTO) is the only international organisation dealing with the global rules of trade between nations. Its main function is to ensure that trade flows as smoothly, predictably and freely as possible so that consumers and producers can enjoy secure supplies and greater choice of the finished products, components, raw materials and services that they use. Producers and exporters may have the information that markets will remain open to them to ensure the existence of a more prosperous, peaceful and accountable economic world.

World Trade Organisation has multilateral trading system comprising agreements, negotiated and signed by governments of

* Lecturer in the Department of English, D.A.V. Degree College, Varanasi.

a large majority of the world's trading nations with the purpose of helping producers, exporters and importers conduct their business. These signed agreements and contracts provide the legal ground-rules for international commerce ensuring important trade rights of member countries thereby binding governments to put curbs on their trade policies to the benefit of everyone.

World Trade Organisation's increased emphasis and environmental policies is relatively a recent phenomenon. WTO, in its Preamble, has stated the need and objective of protecting and preserving the environment. To quote from the Preamble, "Recognizing that their relations in the field of trade and economic endeavour should be conducted with a view to raising standards of living, ensuring full employment and a large and steadily growing volumes of real income and effective demand and expending the production and trade in goods and services while allowing for the optional use of the world's resources in accordance with the objective of sustainable development, seeking both to protect and preserve the environment and enhance the means of doing so in a manner consistent with their respective needs and concerns at different levels of economic development."

WTO has no specific separate Agreement dealing with the environmental protection, however, there are a number of WTO Agreements which include provisions dealing with environmental concerns. WTO provisions relevant to trade-related environmental issues include the 'Non-discrimination Principles of Most-favoured-nation and National Treatment, certain sections of GATT Article 20(i.e.xx) Environmental Concerns are also addressed in a member of different WTO Agreements, for instance the Agriculture Agreement etc.

The relevant provisions concerned with environmental issues are :

1. Non-discrimination
2. Goods: general exceptions of GATT Article 20(XX)
3. Services: general exceptions of GATS Article 14(XIV)
4. Technical Barriers to Trade (TBT) Agreement

5. Sanitary and Phytosanitary Measures (SPS) Agreement
6. Intellectual Property (TRIPS) Agreement
7. Agriculture Agreement

1. Non-discrimination Provision has two principles one is MFN (Article-1) i.e. Most-favoured-nation treatment that means all member countries are on an equal footing and all share the benefits of any moves towards lower trade barriers. A WTO member cannot treat a product of another country more favourable than the products of other WTO members (except in certain circumstances such as under regional free-trade agreements or preferential treatment for developing countries exports). The MFN principle ensures that developing countries and others with little economic leverage are able to benefit freely from the best trading conditions whenever and wherever they are negotiated. This shows that a few powerful nations cannot discriminate between "like" products imported from different trading partners. The other criterion is National Treatment that entails equal treatment for foreign and domestic goods and services. Article III of GATT stipulates that once goods have entered a market, they must be treated on equal terms with equivalent domestically produced goods. Article XVII of GATS, Article 3 of TRIPS also specify this principle.

Non-discrimination is the main principle on which the rules of the multilateral trading system are founded. This principle prevents the abuse of environmental policies and their use as protectionism in disguise. It ensures that national environmental protection policies cannot arbitrarily discriminate between foreign and domestically made products, or between products imported, from different trading partners.

2. Goods: General Exceptions of GATT (Article-20)

Negotiated as early as 1947, GATT Article XX on "General Exception" lays out a number of specific instances in which WTO members (originally GATT "contracting parties"), may be excepted from GATT rules. These include two sets of circumstances for environmental protection.

The Articles states that:

"Subject to the requirement that such measures are not applied in a manner which would constitute a means of arbitrary or unjustifiable discrimination between countries where the same conditions prevail, or a disguised restriction on international trade nothing in this Agreement shall be construed to prevent the adoption or enforcement by any contracting party of measures:

"... (b) necessary to protect human, animal or plant life or health;

"... (g) relating to the conservation of exhaustible natural resources if such measures are made effective in conjunction with restrictions on domestic production or consumption; ..."

Paragraphs (b) and (g) are designed to allow WTO members to adopt policy measures that would normally be inconsistent with GATT, when "necessary" to protect human, animal or plant life or health (which together can be taken to mean "environment"), or if related to the conservation of exhaustible natural resources.

However, the opening paragraph (the "chapeau") of Article XX is designed to ensure that the GATT—inconsistent measures do not result in arbitrary or unjustifiable discrimination and do not constitute disguised protectionism.

Service: The General Exceptions of GATS Article XIV (14)

Negotiated during the 1986-94 Uruguay Round, the General Agreement on Trade in Services (GATS) contains a "general exceptions" clause, Article XIV, similar to GATT Article XX.

The GATS articles starts with an introduction ("chapeau") that is identical to that of GATT Article XX.

Addressing environmental concerns, paragraph (b) allows WTO members to adopt policy measures that would normally be inconsistent with GATS if this "necessary to protect human, animal or plant life or health' (identical to GATT Article XX (b)).

As under GATT, this must not result in arbitrary or unjustifiable discrimination and must not constitute protectionism in disguise.

The Technical Barriers to Trade (TBT) Agreement

The WTO Agreement on Technical Barriers to Trade seeks to ensure that technical regulations and standards, and their testing and certification procedures, do not create unnecessary obstacles to trade.

In its *preamble*, the agreement recognizes countries' rights to adopt such measures to the extent they consider appropriate—for example, to protect human, animal or plant life or health, or the environment.

Moreover, members are allowed to take measures to ensure that their standards of protection are met. (This is known as adopting "conformity assessment procedures".)

Among the agreement's important features are:

- *non-discrimination* in the preparation, adoption and application of technical regulations, standards, and conformity assessment procedures;
- avoiding *unnecessary obstacles* to trade;
- adopting *international standards* as far as possible;
- the *transparency* of these measures, through governments notifying them to the WTO Secretariat and establishing national enquiry points.

The agreement allows countries to adopt technical regulations, standards and conformity assessment procedures for environmental protection, provided non-discrimination, transparency and other requirements are met.

Sanitary and Phytosanitary Measures (SPS) Agreement

The WTO Agreement on Sanitary and Phytosanitary Measures deals with food safety, and human, animal and plant health and safety regulations.

It recognizes members' rights to adopt SPS measures but stipulates that they must be based on science, should not create unnecessary obstacles to trade, and should not arbitrarily or unjustifiably discriminate between members where similar conditions prevail.

The agreement encourages members to adapt their SPS measures to the areas (regions, countries or parts of countries) that supply their imports.

The SPS Agreement complements the Technical Barriers to Trade Agreement. It allows members to adopt SPS measures for environmental purpose, but subject to such requirements as risk assessment, non-discrimination and transparency.

Intellectual Property (TRIPS) Agreement

The WTO Agreement on Trade-Related Aspects of intellectual Property Rights (TRIPS) refers explicitly to the environment in Section 5, which deals with patents.

It says (in paragraphs 2 and 3 of Article 27—Arts 27.2 and 27.3 for short—of Section 5) that members can make certain inventions ineligible for patenting:

- *To protect human, animal or plant life or health, to avoid serious harm to the environment*: A member can exclude an invention from patentability if it believes the invention has to be prevented (within its territory) for these and certain other objectives.
- *Plants and animals*: Micro-organisms have to be eligible for patenting. So do non-biological and microbiological processes for the production of plants or animals. And invented plant varieties have to be eligible for protection either by patenting, or by an effective system specially created for the purpose ("*sui generis*"), or a combination of the two. Otherwise, plants and animals do not have to be eligible for patenting.

The provisions are designed to address the environmental concerns related to intellectual property protection.

The TRIPS Agreement allows members to refuse to patent inventions that may endanger the environment (provided their commercial exploitation is prohibited as a necessary condition for the protection of the environment). For ethical or other reasons, they can also exclude plants or animals from patentability, subject to the conditions described above.

The Agriculture Agreement

Adopted during the 1986-94 Uruguay Round, the WTO Agriculture Agreement seeks to reform trade in agricultural products, and provide a basis for market- oriental policies.

In its preamble, the agreement reiterates members' commitment to reform agriculture in a manner that protects the environment.

Under the agreement, domestic support measures with minimal impact on trade (known as "green box" policies) are allowed and are excluded from reduction commitments—they are listed in Annex 2 of the Agreement. Among them are expenditures under environmental programmes, provided that they meet certain conditions. Again, the exemption enables governments to capture "positive environmental externalities".

Relevant Decision

Two ministerial decisions addressing environmental issues were adopted at the end of the Uruguay Round.

A ministerial *Decision on Trade and Environment,* created the Committee on Trade and Environment (CTE) with the aim of making international trade and environmental policies support each other. The decision contains the work programme of the CTE.

Ministers also adopted a *Decision on Trade in Services and the Environment.* It instructs the CTE to examine and report on the relationship between services trade and the environment, including the issue of sustainable development, of *GATS Article XIV* are required. The CTE has taken up this issue as part of its work programme.

The committee's work is based on two important principles:

- WTO is only competent to deal main trade. In other words, in environmental issues its only task is to shady questions that arise when environmental policies have a significant impact on trade. The WTO is not an environmental agency. Its members do not work it to intervene in national or international environmental policies or to set environmental standards. Other agencies that specialize in environmental issues are better qualified to undertakes those tasks.

- If the committee does identify problems, its solutions must continue to uphold the principles of the WTO trading system.
- WTO members are convinced that an open, equitable and non-discriminatory multilateral trading system has a key contribution to make to national and international efforts to better protect and conserve environmental resources and promote sustainable Development. This was recognised in the results of the 1992 UN conference on Environment and Development in Rio (the "Earth Summit") and its 2002 Successes, the world Summit on Sustainable Development in Johannesburg.

REFERENCES

1. www.wto.org
2. *Yojana* Vol. 48, Nov. June 2004.

24

Municipal Solid Wastes

Some Important Issues

Dr. Devendra Mohan*

All the wastes arising from any human and animal activities and which are normally solid or near solid in state are referred to as solid wastes. The term refuse is used interchangeably with the phrase, 'solid wastes' (Peavy, Rowe and Tchobanoglous, 1986; Tchobanoglous *et al.*, 1977). Broad categories of solid wastes are as mentioned below:

(i) Municipal Solid Wastes (MSW)

(ii) Industrial Solid Wastes

(iii) Hazardous Solid Wastes

Industrial wastes originate from industrial activities and include rubbish, demolition and construction wastes, hazardous wastes, special types of wastes *etc*.. Hazardous wastes are those, which pose a substantial danger immediately or over a period of time to human, plants or animal life. These have the characteristics like ignitability, corrosivity, reactivity or toxicity. Chemicals, radioactive materials, biological wastes, flammable matter and explosives are considered as hazardous wastes.

* Reader, Environmental Engineering Division, Department of Civil Engineering, Institute of Technology, Banaras Hindu Unitersity, Varanasi.

Municipal Solid Wastes (MSW)

Municipal solid wastes (MSW) may be classified as wastes, rubbish, ashes and residues, demolition and construction wastes, special wastes (street sweepings, roadside-litter, catch-basin debris, dead animals, abandoned vehicles *etc.*) and treatment- plant wastes (solid and semi-solid wastes from water, waste water and industrial waste treatment facilities).

Rubbish is also a term, which includes combustible and noncombustible solid wastes, excluding food wastes and other putrescible materials. It consists of combustible materials like paper, cardboard, plastics, textiles, rubber, leather, wood, garden-trimmings *etc.*. Non-combustible items like glass, crockery, tin and aluminium cans, ferrous and non-ferrous metals, dirt and construction wastes also make a part of rubbish (Peavy, Rowe and Tchobanoglous, 1986).

Important sources of MSW are residential, commercial, treatment plant sites and from open areas like parks *etc.*, which vary largely in size characteristics (Winkler and Wilson, 1973). In Indian conditions, 800-1000 gm/day in Delhi and 300-400 gm/day of MSW is generated per capita. In a typical Indian city, about 75 per cent of MSW consists of organic wastes, which can be subjected to generation of an important green fuel like methane (Mishra, 2005) by application of the process of anaerobic digestion. Functional elements of a solid waste management system may be enumerated (Tchobanologous *et al.*, 1977) as mentioned below:

(i) Waste generation

(ii) On-site handling, storage and processing

(iii) Collection

(iv) Transfer and transport

(v) Processing and resource-recovery

(vi) Disposal

Chemical Composition of MSW

Representative data on ultimate analysis of typical MSW are given in Table 24.2 (Tchobanoglous *et al.*, 1977). Using the data given in Table 24.2, an empirical formula for biodegradable fraction of MSW can be derived and thus, any MSW can always be expressed as a hypothetical chemical compound. These empirical formula may be described in two different forms of general expressions as $C_xH_yO_zN_aS$ or $C_bH_dO_eN$, where the subscripts (x, y, z, a) or (b, d, e) have various values under different conditions of composition. From these expressions, approximate energy values can be calculated using modified Dulong formula as represented below by equation (1):

$$\text{Energy, kJ/kg} = 337C + 1428(H - O/8) + 95S \quad \text{.....................} (1)$$

Where,

C = Carbon (%)

H = Hydrogen (%)

O = Oxygen (%)

S = Sulphur (%)

Green Energy from MSW

Typical data on energy and chemical contents, and inert residue of MSW may be as presented in Tables 24.1 and 24.2 (Peavy, Rowe and Tchobanoglous, 1986). Energy values may be reported on 'as discarded basis', 'dry basis' or on 'ash-free dry basis'. These energy values are interrelated by the equations (2 and 3) as given below:

kJ/kg (dry basis) = kJ/kg (as discarded) x 100/(100-% moisture) .... (2)

kJ/kg (ash-free dry basis) = kJ/kg (as discarded) x 100/(100- % moisture-% ash content) .... (3)

A large fraction of MSW being organic, the potential for generation of green energy like methane is quite high, when MSW is subjected to anaerobic decomposition. In addition to methane, a good amount of digested (stabilized) sludge is obtained, which can be used as good manure after small treatment procedure applied.

There are two very important cycles in nature involving the growth and decay of organic matter, *i.e.*, aerobic cycle, which uses oxygen in the process of decay of organic matter, and anaerobic cycle that employs the conditions involving the absence of oxygen for decay of the biodegradable materials (Wright and Nebel, 2002). Therefore, MSW can also be stabilized by applying aerobic or anaerobic digestion processes.

Table 24.1: Typical Energy Contents of Certain Components of MSW

Component	*Inert residue (%)*		*Energy, kJ/kg*	
	Range	*Typical*	*Range*	*Typical*
Food wastes	2-8	5	3,500-7,000	4,650
Paper	4-8	6	11,600-18,600	16750
Cardboard	3-6	5	13,950-17,450	16,300
Plastics	6-20	10	27,900-37,200	32,600
Textiles	2-4	2.5	15,100-18,600	17,450
Rubber	8-20	10	20,900-27,900	23,250
Leather	8-20	10	15,100-19,800	17,450
Garden trimmings	2-6	4.5	2,300-18,600	6,500
Wood	0.6-2	1.5	17,450-19,800	18,600
Miscellaneous organics	2-8	6	11,000-26,000	18,000
Glass	96-99	98	100-250	150
Tin cans	96-99	98	250-1,200	700
Nonferrous metals	90-99	96	–	–
Ferrous metals	94-99	98	250-1,200	700
Dirt, ash, bricks *etc.*	60-80	70	2,300-11,650	7,000
MSW as a whole	–	–	9,300-12,800	10,500

Aerobic digestion process is essentially a continuation of aeration process, with the volume being reduced by thickening in secondary clarifier and sludge thickener. Aerobic digestion is an energy consumptive (endogenous respiration) process, in which the microorganisms are forced to metabolize their own protoplasm and this process is not as sensitive as is anaerobic digestion, to various environmental factors. As a consequence, this process is not as easily upset by external environment as occurs in case of the anaerobic counterpart.

Table 24.2: Chemical Composition of Important Components of MSW

Component	*Composition (dry basis), %*					
	Carbon	*Hydrogen*	*Oxygen*	*Nitrogen*	*Sulfur*	*Ash*
Food wastes	48.0	6.4	37.6	2.6	0.4	5.0
Paper	43.5	6.0	44.0	0.3	0.2	6.0
Cardboard	44.0	5.9	44.6	0.3	0.2	5.0
Plastic	60.0	7.2	22.8	–	–	10.0
Textiles	55.0	6.6	31.2	4.6	0.15	2.5
Rubber	78.0	10.0	–	2.0	–	10.0
Leather	60.0	8.0	11.6	10.0	0.4	10.0
Garden trimmings	47.8	6.0	38.0	3.4	0.3	4.5
Wood	49.5	6.0	42.7	0.2	0.1	1.5
Miscellaneous organics	48.5	6.5	37.5	2.2	0.3	5.0
Dirt, ash, bricks *etc.*	26.3	3.0	2.0	0.5	0.2	68.0

The sludge produced in aerobic process is relatively inert, but dewaters poorly and it is often disposed off in a rather dilute state. On the other end, anaerobic digestion offers the following major advantages over the aerobic process (CPHEEO, 1993):

(a) Anaerobic process is a net energy producer. Methane is obtained as a valuable by-product, which is an important source of clean energy.

(b) Application of the sludge obtained can improve the fertility and texture of soils.

(c) Pathogens in the sludge die off during the relatively long detention periods and adverse conditions encountered in anaerobic digestion.

The following are the major disadvantages of anaerobic digestion process (CPHEEO, 1993):

(a) Relatively large and closed (air-tight) digestion tanks are required, which need high capital investment.

(b) Strict process control and monitoring are needed due to higher sensitivity of the microorganisms to changes in environmental factors.

(c) Supernatant has higher oxygen demand along with more concentration of nitrogen-bearing compounds and suspended solids.

In most of the biological treatment systems, three processes of biomass synthesis, energy production and endogenous respiration at cellular level take place almost simultaneously (Metcalf and Eddy, 1979). Since anaerobic process can yield methane, this process is being promoted for digestion of wastewater treatment plant sludges (whose important characteristics are as given in Table 24.3), and also for decomposition of biomass for generation of CH_4. The main objectives of the sludge treatment are as follows (Barnes *et al.*, 1981; Fair and Geyer, 1957):

(a) Stabilization of biodegradable matter

(b) Reduction in volume of the sludge

(c) Destruction of pathogens

(d) Collection of valuable by-products

(e) Disposal of the end product in a safe and aesthetically acceptable manner

Table 24.3: Important Characteristics of Sludge from a Domestic Sewage Treatment Plant

Sludge type	*Moisture content (% weight)*	*Organic content (% dry weight)*
Primary	93-97	40-80
Activated	98-99.5	65-75
Digested	96-99	30-60
Humus	94-99	65-75
From high rate plastic medium	92-97	65-75

Incineration, pyrolysis and composting processes have also been applied traditionally for safe disposal of the sludge produced in municipal sewage treatment plants, and also for producing thermal energy to some extent. These processes are being described below in very brief. Nowadays, anaerobic process is being seriously considered for application on a large scale as it can generate clean energy along with achievement of very safe disposal of the sludges and municipal solid wastes. Various processes commonly involved in the sludge treatment and disposal are listed in Table 24.4.

Incineration

Incineration can be used for sludges, which contain very little water and also have very small organic content. Stringent controls to prevent secondary pollution are required for any liquid or gaseous discharge from the incinerators. Sludge solids have calorific values similar to conventional fossil fuels and dry sludges can be burnt without any additional fuel consumption. Two types of furnaces employed for the purpose are multiple hearths and fluidized bed furnaces.

Pyrolysis

Pyrolysis refers to the process of combustion under condition of reduced supply of oxygen and therefore, complete oxidation of the waste is not possible. Partially-combusted gaseous products from low-temperature pyrolysis unit are passed to high temperature furnace, where complete oxidation is carried out. Usually, additional fuel is required to pyrolyze partially wet sludges of low calorific value. Municipal refuse may provide an economic source of fuel. Pyrolysis of municipal refuse-derived fuel and dewatered municipal wastewater sludges can provide a system, which can yield thermal energy in addition to facilitating their safe and aesthetic disposal.

Table 24.4: Treatment and Disposal Options for Wastewater Sludge (Barnes *et al.*, 1981)

Thickening	*Stabilization*	*Dewatering*	*Partial Disposal*	*Ultimate Disposal*
Gravity	Anaerobic digestion	Drying beds	Incineration	Sanitary landfill
Floatation	Aerobic digestion	Filter press	Pyrolysis	Crop land application
Centrifugation	Lagooning	Centrifugation	Wet air oxidation	Sea/ocean
Elutriation	Heat treatment	Vacuum filtration	Composting	
		Belt press lagooning		

Aerobic Treatment

Composting of dewatered municipal sludges may be carried out along with refuse to produce agricultural or horticultural compost. This process is applied after separation of components (undesirable for this process) from municipal solid wastes. The stable product is quite acceptable, if industrial inputs to garbage and wastewater are not significant.

Biodegradation

Microorganisms utilizing organic materials for energy are known as heterotrophs, while those using inorganic materials for the same purpose are called as autotrophs. Autotrophs using inorganic chemicals for energy production are called chemotrophs and on the other end, those utilizing sunlight for energy generation are termed as phototrophs (Schlegel, 1989).

In the life processes, the production of energy from release of electrons does not occur automatically, but through a series of steps, which produce a specific compound (Adenosine triphosphate: ATP) with a very high energy content. The electrons, from both hydrogen and the substrate, emerge from the system to reduce a final external electron acceptor. These acceptors may be one of the following:

- For aerobic processes, it is oxygen.
- For anaerobic processes, these may be NO_3^-, SO_4^{--} and CO_2. When the acceptor is NO_3^-, the system is known as anoxic.

The reaction involving sulfate reduction and methane fermentation release the least amount of energy in comparison to all other biochemical reactions of this nature. These reactions can be carried out only by specific group of anaerobic bacteria and cannot proceed in presence of oxygen and/or nitrates (Sincero and Sincero, 1999). Anaerobic systems in wastewater treatment can be applied for treating organic sludges removed from the sewage in primary and secondary sedimentation, following aerobic treatment. There are a number of simple ways through which this process has been applied for the sludge treatment, *e.g.*, anaerobic ponds, septic tanks, imhoff tanks, anaerobic-aerobic lagoons and the similar systems (Barnes *et al.*, 1981).

Most of the conventional anaerobic processes have essentially been suspended growth systems. Afterwards, systems using fixed film processes like anaerobic filters were developed, which can attain effective treatment of both dilute and concentrated wastewaters, even at normal temperature. In these filters, a long mean cell residence time (about 100 days) can be maintained for mean hydraulic detention time of only a few hours. Later on, taking into consideration the clean energy content of the product of anaerobic decomposition like methane, it was thought that the process could be successfully and effectively applied for generation of valuable energy-bearing gas from solid wastes, biomass, droppings of domesticated animals, human excreta and any other biodegradable materials of this type.

As in case of biological combustion processes, the reaction occurring in anaerobic process may or may not attain completion. Decomposition of solid wastes in sanitary landfills, digestion of sludge in a domestic and/or industrial sewage treatment plant(s) and generation of fuel gas from biomass are based on the chemistry of anaerobic processes.

Any organic waste or biomass can be expressed using an empirical chemical formula, which can always be derived taking different constituents of the waste/biomass into account (Peavy, Rowe and Tchobanoglous, 1986). When an organic compound represented by a generalized empirical chemical formula as $C_xH_yO_zN_aS_b$ undergoes microbiological decomposition or digestion, carbon dioxide is invariably generated irrespective of the fact whether the process is aerobic or anaerobic. If there is no intentional addition of nitrates or sulfates, the electron acceptor will naturally be CO_2. For complete anaerobic reaction, equations (4 and 5) may be derived:

Donor Half-reaction

$$[1/(y+4x-3a-2z-2b)]\, C_xH_yO_zN_aS_b + [(2x+a-z)/(y+4x-3a-2z-2b)]\, H_2O \longrightarrow [(x-a)/(y+4x-3a-2z-2b)]\, CO_2 + [a/(y+4x-3a-2z-2b)]\, NH_4^+ + [a/(y+4x-3a-2z-2b)]\, HCO_3^- + [b/(y+4x-3a-2z-2b)]\, H_2S + H^+ + e^- \quad (4)$$

Acceptor Half-reaction

$$(1/8)\, CO_2 + H^+ + e^- \longrightarrow (1/8)\, CH_4 + (1/4)\, H_2O \quad (5)$$

Combination of half-reactions shown by equations (4 and 5) gives equation (6) as shown below, which represents the complete reaction for anaerobic process:

$$[1/(y+4x-3a-2z-2b)]\, C_xH_yO_zN_aS_b + [(4x+7a+2b-2z-y)/4(y+4x-3a-2z-2b)]\, H_2O \longrightarrow [(4x+2z+2b-5a-y)/8(y+4x-3a-2z-2b)]\, CO_2 + (1/8)\, CH_4 + [a/(y+4x-3a-2z-2b)]\, NH_4^+ + [a/(y+4x-3a-2z-2b)]\, HCO_3^- + [b/(y+4x-3a-2z-2b)]\, H_2S \quad (6)$$

In case of incomplete oxidation of the substrate in anaerobic decomposition, if $C_fHgO_hN_kS_m$ represents the residue molecule, equation for the mass of methane produced may be derived using the same principles as applied for the case discussed above. The fuel gas generated may be calculated using equation (7) as given below:

$$(\text{Mass of methane})/(\text{Mass of } C_xH_yO_zN_aS_b) = \{[(4x + y - 3a - 2z - 2b) - Y_{dc}(4f + g - 3k - 2h - 2m)]/8\} \; CH_{4/}C_xH_yO_zN_aS_b \quad (7)$$

Where, Y_{dc} represents the cell yield under anaerobic conditions, in moles of $C_fHgO_hN_kS_m$ per mole of $C_xH_yO_zN_aS_b$, which is the electron donor.

Methane fermentation is important in environmental engineering, as it is the major process by which the wastes can be stabilized in anaerobic digestion. The same process can be effectively used for producing fuel gas from biomass also. Increased interest in this process has been shown in recent years in face of dwindling supply of energy resources. About 70 per cent of methane produced from complex waste/biomass results from fermentation of acetic acid, which is formed as an intermediate compound in anaerobic degradation of carbohydrates, proteins, fats, and other organic compounds (Jeris and McCarty, 1965).

In anaerobic digestion of carbohydrates, an external electron acceptor is not required as the organic matter itself acts as electron acceptor. Many other compounds can serve as terminal electron acceptor in biological processes. Most important amongst these are ferric iron and iron-containing organic compounds, oxidized sulfur compounds and even halogenated organic compounds. Oxidation-reduction reactions with iron are particularly important in groundwater conditions.

Nitrite can also serve as an electron acceptor, though it is rarely found in significant concentrations in environment (Sawyer, McCarty and Parkin, 1994). In this process, two groups of bacteria, *e.g.*, acid formers and methane formers, work in harmony. Acid forming microorganisms carry the degradation to the acid stage and then, methane forming bacteria complete the conversion into methane and carbon dioxide. Very simplified representation of generalized expressions for aerobic (equation 8) and anaerobic (equation 9) processes, taking place on the same substrate (*e.g.*, with empirical formula as $C_kH_mO_nN_qS_uP_v$) can be given as shown below for the sake of comparison:

$$C_kH_mO_nN_qS_uP_v + O_2 \longrightarrow CO_2 + H_2O + NO_3^- + PO_4^{3-} + SO_4^{2-} + \text{synthesized new cells} + \text{energy} \quad (8)$$

$$C_kH_mO_nN_qS_uP_v + NO_3^- + PO_4^{3-} + SO_4^{2-} \longrightarrow CO_2 + CH_4 + N_2 + PH_3 + H_2S + \text{synthesized new cells} + \text{energy} \quad (9)$$

The products of anaerobic process are usually more objectionable than those of aerobic process. Methane and phosphine are combustible gases, while hydrogen sulfide and other products, like mercaptans and amines, are malodorous (Barnes *et al.*, 1981).

For design of the anaerobic reactor, an environmental engineer needs to know all the transformations taking place in the biological process. Portion of electron donor, converted for energy need, should be known so that the quantities of oxygen, nitrates or sulfates required may be found out, and the quantity of methane produced may also be determined. To determine the quantities of nutrients like nitrogen and phosphorus for production of microbial cells, portion of the electron donor converted to cell mass must also be known. The cell mass-produced should also be known for designing the sludge handling facilities (Sawyer, McCarty and Parkin, 1994).

Mass balance for the system can be performed by using a balanced chemical reaction for overall biological conversion, which involves proper terms for synthesis and energy production at the cellular level. An empirical formula, $C_5H_7O_2N$, is commonly used, while considering the synthesis of the bacterial cells. A balanced overall reaction for anaerobic decomposition may be written as shown in equation (10) below (McCarty, 1975):

$$R = f_s R_c + f_e R_a - R_d \quad (10)$$

Where,

R stands for overall biological conversion

R_c represents the half reaction for synthesis of bacterial cells,

R_a represents the half reaction for the electron acceptor, and

R_d represents the half reaction for the electron donor.

f_s and f_e represent the portions of the electron donor for cell synthesis and energy production respectively.

Therefore,

$f_s + f_e = 1$ -- (11)

Methane Production

Microbiological Aspects

Methane producers live in anaerobic environments having plenty of organic matter, *e.g.*, rumen and intestine system of animals, freshwater and marine sediments, swamps and marshes, hot springs, and even within anaerobic protozoa. Methanogens obtain energy by converting CO_2, H_2, formate, methanol, acetate and other such compounds to either methane, or methane and CO_2. They behave as autotrophic, when thriving on H_2 and CO_2. This is the largest group of archaeobacteria, encompassing at least three orders and thirteen genera, which differ greatly in overall shape, 16S RNA sequence, cell wall chemistry and structure, membrane lipids and many other features (Prescott, Harley and Klein, 1990). These microorganisms can produce methane anaerobically due to their unusual metabolic processes. They contain several unique cofactors like tetrahydromethanopterin, methanofuran, 2-mercaptoethanesulfonic acid, coenzyme F_{420} and coenzyme F_{430}. It seems that ATP synthesis is linked with methanogenesis by electron transport, proton pumping and a chemiosmotic mechanism.

In the form of a cause to one of the major environmental problems, methane has shown to be an effective green house gas because of its capability to absorb infrared radiation. Atmospheric methane concentrations are rising over the last 200 years and it can significantly promote future global warming. It is recently found that methane formers can oxidize Fe^0 and utilize it to produce methane and energy, thus significantly contributing to corrosion of buried or submerged pipes and other objects (Prescott, Harley and Klein, 1990). The conversion of organic matter is supposed to occur in two or three steps under anaerobic conditions. In three step sequence, the steps involved are as mentioned below:

(i) The first step involves enzyme-mediated transformation (liquefaction) of higher molecular weight compounds into those, which are suitable for cell synthesis and energy production.

(ii) The second step involves the bacterial conversion of compounds formed in the first step into identifiable low molecular weight compounds.

(iii) The third step involves bacterial conversion of intermediate compounds into simpler end products, which are mainly methane and carbon dioxide.

In the two-step process, the first two steps as described above are thought to occur simultaneously and are considered together as the first step. In two-step sequence, the microbes involved are divided into two broad groups, *i.e.*, the first group forming mainly acetic and propionic acids is also described as nonmethanogenic and it consists of facultative and obligate anaerobes like Clostridium spp., Peptococcus anaerobus, Bifidobacterium spp., Desulphovibrio spp., Corynebacterium spp., Lactobacillus, actinomycetes, Staphylococcus and Escherichia coli. Other physiological groups present under these conditions include the microorganisms, which produce proteolytic, lipolytic, ureolytic or cellulytic enzymes (Higgins and Burns, 1975). Various species of Pseudomonas, Flavobacterium, Alcaligenes, Escherichia and Aerobacter contribute to the acids production in anaerobic degradation of substrates (McKinney, 1962).

The principal genera of microbes responsible for the second step of anaerobic decomposition are similar to those found in stomachs of ruminant animals and in organic sediments collected from lakes and rivers as already mentioned. Important of these microorganisms are in the shape of rods (Methanobacterium and Methanobacillus) and spheres (Methanococcus and Methanosarcina) (Balch *et al.*, 1979; Jones, Nagle and Whitman, 1987; McKinney, 1962; Metcalf and Eddy, 1979). The ease in growth of the facultative bacteria give them an edge over the strict anaerobes much in the same way in which they maintain their predominance over the strict aerobes in activated sludge systems. It is interesting to note that certain bacteria in one phase of anaerobic digestion and those predominant in activated sludge systems belong to one and the same group.

These facultative and anaerobic bacteria of wide variety hydrolyze and convert the complex materials to low-molecular weight compounds, important amongst which are short chain fatty acids like acetic, propionic, butyric and to a lesser extent, isobutyric, valeric, isovaleric and caproic acids. An accumulation of such acids may lead to failure of the treatment process, if buffering capacity of the system is exceeded and consequently, pH falls to unfavorable levels. Under favorable operating conditions, volatile acids content of digested sludges may be in the range of 50–250 mg/l, expressed as acetic acid.

Methane formers are ubiquitous in nature. But, their population is very small in comparison to that of many acid-formers. This disparity in numbers is the main reason for the troubles, which are encountered in starting the units based on anaerobic treatment process, without seeding. Raw sewage sludges and industrial wastewaters have many times quite low buffering capacities, which are soon spent and free acids start accumulating, thus resulting into continued depression of pH of the system. Fortunately, the alkalinity of the digesting sludges is naturally high because of the solubilization of CO_2 produced by the process itself, and its subsequent conversion to HCO_3^-. When digestion is proceeding satisfactorily, the alkalinity normally ranges from 1000 to 5000 mg/l and volatile acids will be less than 250 mg/l.

At pH values below 6.5 (complete end of the methanogenic activity is observed below pH of 6.2), methane formers are seriously inhibited, but not acid formers until pH levels fall to about 5. Under such conditions, volatile acids may accumulate up to levels ranging from 2000–6000 mg/l or sometimes even more, depending upon the biodegradable solids content of the sludge. To provide buffering under such conditions, dilution or addition of neutralizing agents such as lime need be done. Therefore, volatile acids determination along with pH measurement becomes very important in controlling the environmental conditions during the beginning and progress of the anaerobic treatment.

Chemical Aspects

With regard to the specific mechanisms involved in the formation of methane, it appears that two pathways are possible, depending on the nature of starting substrate (Zinder, 1984). The bacteria involved seem to be capable of utilizing the following three types of substrates:

(a) The lower fatty acids containing six or less carbon atoms (formic, acetic, propionic, butyric, valeric and caproic)

(b) The normal and isoalcohols containing one to five carbon atoms (methanol, ethanol, propanol, butanol and pentanol)

(c) Three inorganic gases (hydrogen, carbon monoxide and carbon dioxide)

The two mechanisms involved may be explained using the following equations (12-20). In the first one, which involves substrates like ethanol, butyrate and hydrogen, methane is produced from substrate oxidation and from reduction of atmospheric carbon dioxide. In the second mechanism, which involves substrates such as acetate and propionate, methane is generated by reduction of carbon dioxide, which is formed during the oxidation of substrates.

(i) Reduction of atmospheric carbon dioxide:

$$2C_2H_5OH + CO_2 \longrightarrow 2CH_3COOH + CH_4 \quad \text{------} \quad (12)$$

$$4H_2 + CO_2 \longrightarrow 2H_2O + CH_4 \quad \text{------} \quad (13)$$

(ii) Reduction of carbon dioxide formed in the reaction:

$$CO + H_2O \longrightarrow CO_2 + H_2 \quad \text{------} \quad (14)$$

$$CO_2 + 4H_2 \longrightarrow 2H_2O + CH_4 \quad \text{------} \quad (15)$$

$$CO + 2H_2 \longrightarrow H_2O + CH_4 \quad \text{------} \quad (16)$$

$$4C_2H_5COOH + 8H_2O \longrightarrow 4CH_3COOH + 4CO_2 + 24H \quad \text{------} \quad (17)$$

$$3CO_2 + 24H \longrightarrow 6H_2O + 3CH_4 \quad \text{------} \quad (18)$$

$$4C_2H_5COOH + 2H_2O \longrightarrow 4CH_3COOH + 3CH_4 + CO_2 \quad \text{------} \quad (19)$$

$$CH_3COOH + CO_2 \longrightarrow CO_2 + CH_4 \quad \text{------} \quad (20)$$

In addition, many other groups of bacteria like denitrifiers and desulphovibrio carry out the reduction processes of nitrates and sulfates under anaerobic conditions.

The Reactor and Gaseous Products

Reactor for methane generation (by anaerobic digestion of MSW) consists of closed tanks with airtight cover. For the purpose of design, a completely mixed and continuous flow model without solids recycle is usually considered, although the first two conditions seldom apply exactly. Treatment plants processing relatively less volume of sewage utilize a type of reactor, commonly called as standard-rate anaerobic digester, while those handling huge volumes (normally more than 4000 m^3/day) of wastewater employ high rate digesters. Due to stratification, the single stage or standard-rate or cold anaerobic digesters develop four distinct zones as mentioned below (Benefield and Randall, 1980; Hammer and Hammer, Jr., 1998):

(a) Scum zone, where floating materials tend to accumulate.

(b) Supernatant liquor zone, where the water released from digesting solids get accumulated.

(c) Active digestion zone, where mainly anaerobic degradation process occurs.

(d) Digested sludge zone, where stabilized solids settle down for ultimate removal after the designed time interval.

Product gases from this process mostly consist of 65 to 70% methane, 25 to 30% CO_2 and other gases in trace amounts. Approximately, one m^3 of total gas is produced per kilogram of the solids digested. Heat content of methane is about 36,000 kJ/m^3. It has also been reported that this methane can provide a significant fraction of energy required to operate the same wastewater treatment plant (Banerjee and O'Connor, 1977). The supernatant liquor has very high amounts of solubilized organics and solids and therefore, it need be recirculated through the plant for further treatment. Solids withdrawn from the conical hopper at the bottom of the well-performing anaerobic reactor should be practically inert and pathogen-free. These may be disposed off safely by various methods.

REFERENCES

Balch, W.E., Fox, G. E., Magrum, L.J., Woese, C.R. and Wolfe, R.S. (1979), Methanogens: Reevaluation of a Unique Biological Group, Mocribiol. Rev., Vol. 43, pp. 260-296.

Banerjee, S. K. and O'Connor, J. T. (1977), Designing More Energy Efficient Wastewater Treatment Plants, Civil Engineering, Vol. 47(7), September Issue, p. 76.

Barnes, D., Bliss, P.J., Gould, B.W. and Valentine, H.R. (1981), Water and Wastewater Engineering Systems, Pitman Press, Bath, U.K.

Benefield, L.D. and Randall, C.W. (1980), Biological Process Design for Wastewater Treatment, Prentice Hall, Inc., Englewood Cliffs, New Jersey, USA.

CPHEEO (1993), Manual on Sewerage and Sewage Treatment, Second Edition, Central Public Health and Environmental Engineering Organization, Ministry of Urban Development, Government of India, New Delhi.

Fair, G.M. and Geyer, J.C. (1957), Elements of Water Supply and Wastewater Disposal, John Wiley & Sons, Inc., New York, USA.

Hammer, M.J. and Hammer, M.J., Jr. (1998), Water and Wastewater Technology, Prentice-Hall of India, New Delhi.

Higgins, I.J. and Burns, R.G. (1975), The Chemistry and Microbiology of Pollution, Academic Press, London, U.K.

Jeris, J.S. and McCarty, P.L. (1965), The Biochemistry of Methane Fermentation Using C^{14} Tracers, Journal of Water Pollution Control Federation, Vol. 37, p. 178.

Jones, W.J., Nagle, D.P. (Jr.) and Whitman, W.B. (1987), Methanogens and Diversity of Archaeobacteria, Microbiol. Rev., Vol. 51, pp. 135-177.

Kaiser, E.R.(1966), Chemical Analysis of Refuse Compounds, Proceedings of National Incineration Conference, American Society of Mechanical Engineers, New York, USA.

McCarty, P. L. (1975), Stoichiometry of Biological Reactions, Progress in Water Technology, Vol. 7, Pergamon, London, U.K.

McKinney, R.E. (1962), Microbiology for Sanitary Engineers, McGraw-Hill Book Company, Inc., New York, USA.

Metcalf and Eddy, Inc. (1979), Wastewater Engineering: Treatment, Disposal and Reuse, Second Edition, Tata McGraw-Hill Publishing Company, New Delhi, India.

Mishra, A. (2005), Environmental Studies: Systems and Solutions, Selective and Scientific Books, New Delhi, pp. 183-186.

Peavy, H. S., Rowe, D.R. and Tchobanoglous, G. (1986), Environmental Engineering, McGraw-Hill Book Company, International Edition, New Delhi, India.

Prescott, L.M., Harley, J.P. and Klein, D.A. (1990), Microbiology, Second Edition, Wm. C. Brown Publishers, Dubuque, Iowa, USA.

Sawyer, C.N., McCarty, P.L. and Parkin, G.F. (1994), Chemistry for Environmental Engineering, Fourth Edition, McGraw-Hill International Editions, Singapore.

Schlegel, H.G. and Bowien, B. (1989), Autotrophic Bacteria, Science Tech Publishers, Madison, Wisconsin, USA.

Sincero, A.P. and Sincero, G.A. (1999), Environmental Engineering: A Design Approach, Prentice Hall of India, New Delhi.

Steel, E.W. and McGhee, T.J. (1979), Water Supply and Sewerage, Fifth Edition, McGraw-Hill International Book Company, International Student Edition, Tokyo, Japan.

Tchobanoglous, G., Theisen, H. and Eliassen, R. (1977), Solid Wastes Engineering: Principles and Management Issue, McGraw-Hill, New York, USA.

Winkler, P.F. and Wilson, D.G. (1973), Size Characterstics of Muncipal Solid Wastes, Compost Science, Vol. 14(5), April Issue.

Wright, R.T. and Nebel, B.J. (2002), Environmental Science: Towards a Sustainable Future, 8th Edition, Prentice-Hall, Inc., New Jersey, USA.

Zinder, S. H. (1984), Microbiology of Anaerobic Conversion of Organic Waste to Methane: Recent Developments, ASM News, Vol. 50(7), pp. 294-298.

25

Pollution Due to Biocides

Dr. Devendra Mohan*

Earlier, fertilizers and other agro-chemicals (applied for plant protection) and irrigation water have been used very liberally and to some extent indiscriminately, to achieve high yield in the process of realizing green revolution. As a consequence, pests and disease-causing organisms have developed resistance to pesticides and other such agro-chemicals, surface and groundwater have become polluted and high levels of chemical- residues are being detected in agricultural products, especially in milk and milk-products, fruits and vegetables.

Pesticides are mostly organic compounds which are applied to control undesirable species of plants and animals. These can also be called as poisons used to bring down the damage to the agricultural production by undesirable life-forms and consequently, to economy. As all these chemicals are used to control (mostly by killing) very large types of unwanted living beings, a proper term to address them may be biocides. At present, more than 10000 types of biocides are used to control damages by insects, weeds/herbs, fungi, bacteria, nematodes, rodents, molluscs, fishes, etc. at various places like agricultural fields, industries and even in homes. Thus, the biocides may be broadly classified as mentioned below:

---

* Reader, Environmental Engineering Division, Department of Civil Engineering, Institute of Technology, Banaras Hindu Unitersity, Varanasi.

(a) Herbicides: Chemicals used to control undesirable plants.

(b) Fungicides: Compounds applied to control various types of fungi.

(c) Insecticides: Used to control insects (commonly called as pests) in agricultural fields.

(d) Nematicides: Employed to inhibit the growth of nematodes.

(e) Rodenticides: Used mainly against rats and mice.

(f) Piscicides: Chemicals used to control unwanted fish species and other aquatic animals.

(g) Molluscicides: Designed to kill molluscs, particularly snails and slugs.

Most of the pesticides used nowadays are synthetic chemicals. Chlorine is mostly found in the most toxic and persistent synthetic chemicals. The chlorine reacts with hydrocarbons to form organo-chlorine compounds, which as a group has the properties of toxicity, persistence and bioaccumulation.

Examples of common biocides may be DDT, 2, 4–D, 2, 4, 5–T, picloram, endosulphan, diquat, paraquat, aldrin, dieldrin, endrin, chlordane, lindane, heptachlor, methoxychlor, heptachlor epoxide, organic phosphates, toxaphene, carbamates, *etc.*. DDT is one of the most widespread molecules amongst all these and it has been banned in developed world much earlier. Similarly, endosulphan and paraquat have been banned in industrialized and developed world for human health reasons, but all these chemicals are still being produced for various uses in different developing countries.

ENVIRONMENTAL IMPACTS

Any type of chronic or acute stress may lead to development of unusual features like accumulation of certain types of chemicals, physical malformations, etc. in exposed plants or animal species. This may also result in reduction in the quality of the related products. Pollution can have adverse effects not only on somatic cells, but also it can cause serious damage even to genetic cells many times. Pesticides, used all over the world, play havoc with environment and human health. India accounts for about one-third of the total pesticide poisoning cases in the world. The problems in developing world are more severe due to indiscriminate use, improper human protection methods employed and careless handling of the pesticides.

Human beings get exposed to such chemicals through formulation, production, packaging, transportation and application on vegetations or animals for various control purposes. There arise the possibilities of contact, inhalation and even accidental ingestion of these toxic chemicals. In addition to exposure through these routes, their entry into food chain of human beings and economic animals occur due to residues in or on raw agricultural commodities or processed food produced there from. Traces of pesticides keep reaching every young or old, healthy and sick, male or female, for long or short period at various stages of life or even throughout the life-span.

Bioconcentration and biomagnification are very peculiar features associated with a large number of the cumulative toxins like pesticides. Biomagnification of these compounds or modified products leads to startling consequences due to progressive and drastic enhancements in concentrations of toxin(s) in various tissues of the life-forms, which are successively situated at higher and higher trophic levels of the food pyramids in any type of eco-system. This situation becomes worst in case of aquatic eco-systems as the food chain involved is relatively long and therefore, the enhancement in concentration is exceptionally high (even sometimes by a million times or more) in the tissues of large fishes or other aquatic animals at or near the top of the food pyramid. Therefore, in present era of widespread pollution, it is safer to be as near to producers (plants, vegetables and trees) as possible so that bioaccumulation affects us adversely to the minimum possible extent. In other words, it can be said that vegetarian diet is the safest diet as it exposes the human beings to the minimum possible levels of various cumulative toxins.

The pollutants find their way into the body of the cattle through food chain and may also multiply in concentration many times (in comparison to their original concentrations in the environment), depending on the length of the food chain upto the level of the particular consumer. Therefore, the animal products used as food like milk, meat, eggs, etc. show much higher concentrations of various cumulative toxins in comparison to their concentrations in the surrounding environments of the animals. Cumulative toxins are those toxic elements (*e.g.*, Cd, Hg, As and Pb) or chemical compounds (many pesticides, herbicides, insecticides, etc. used for various agricultural and other purposes), which have a tendency to accumulate in the body of the exposed animals dose by dose. Therefore, such chemicals may be retained in the body for a long time and may, as a consequence, lead to development of visible physiological disorders after an acute or long chronic exposure(s). D. D. T., 2,4-D, 3, 4, 5-T ,etc. may be quoted as examples of cumulative toxins, which have found widespread use in agriculture for long and have also shown their adverse impacts on animal world by the process of biomagnification through food chain. D.D.T. is a very persistent chemical and it may still be found in environment in original form or as by-products after degradation.

Commercial insecticides like carbofuran and herbicide like butachlor have been evaluated for the rice-field cyanobacterium, Anabaena doliolum, which is an economically important nitrogen-fixing bacterium. It was found that the pesticides caused inhibition to growth of the microorganisms and also, led to reduction in protein and chlorophyll content in various plant tissues and products (Pandey, 2004). Similarly, the fishes reared in polluted surface waters having effluents from carpet manufacturing industries in Bhadohi (recently named as Sant Ravidas Nagar) District of Uttar Pradesh (India) have shown accumulation of certain dyes (used in these industries) in various types of tissues in their bodies.

The study conducted by Lata and Srivastava (2004) has shown that the dyes like Chrome Black T and Chrome Fast Red F tend to accumulate to the maximum extent in kidney and then, followed by brain, liver and gut of the fish. The accumulation of these dyes ultimately leads to damage of the accumulating tissues, causing the death of the fish commonly found in the region and used under experimental observation, *i.e.*, Puntius sophore. Similarly, the crops grown in polluted soil (in field) or using partially treated or untreated sewage for irrigation may suffer from the degradation in quality of the agricultural produce due to accumulation of certain toxic chemicals. A sudden decrease in chlorophyll content in shoot leaves of Vicia faba L. has been reported to result in chlorosis and necrosis in young leaves. Root and/or shoot leaves protein contents under copper stress also showed declination, whereas enzyme peroxidase showed enhancement in concentration as the concentration of exposure to copper increased (Arya, Singh and Ray, 2004).

In aquatic food chain, which starts from phytoplanktons and ultimately ends mostly with fishes or other large aquatic animals at or near the top of the food pyramid, the interconnected links are quite large in number. In other words, it can be said that the trophic levels in food pyramid of aquatic ecosystem are relatively higher in number than those found in other types of ecosystems. Consequently, the process of biomagnification leads to much higher accumulation of cumulative toxins in fishes, crabs and other aquatic animals, which are used as food also by human beings. Certain chemicals have shown carcinogenic, mutagenic, teratogenic potential in addition to other types of physiological effects on experimentally or accidentally exposed mouse and also on other aquatic animals as reported by Yadav and Shukla (2004), Srivastava and Singh (2004) and Pandey and Parveen (2004) to mention a few. Singh (2004) has also reported that certain cumulative toxins may multiply in concentration through food chain, even if treated domestic sewage is used for irrigation.

The process of bioaccumulation depends on many factors like lipid and water partition co-efficient; characteristics of aquatic organisms; environmental conditions like temperature, pH, salinity, etc.; concentration of cumulative toxin(s); duration of exposure; length of food chain; etc.. Organo-chlorine compounds have a greater tendency to bioaccumulate in comparison to organo-phosphorus compounds and carbamates. Because of unsustainable agricultural and industrial practices remaining prevalent for long in the past, air, soil and water have all become polluted to an alarming extent. Table 25.1 shows the percentage of assessed river lengths and lake areas affected by pollution and Table 25.2 presents the number of States in U.S.A. reporting groundwater contamination.

It has also been shown that the crops (particularly leguminous ones) have a high tendency to accumulate heavy toxic metals like zinc (Zn), lead (Pb), cadmium (Cd), nickel (Ni), etc. as reported by

Prakash and Srivastava (2004). In another study conducted on a number of crops and other vegetations like Jowar (Sorghum vulgare), maize (Aea maze), Doob (Cynodon dactylon), Bathua (Chinopodium album) and cow pea (Vigna sinensis), a significant accumulations of Cd, Pb, Cu and Zn have been reported by Gupta (2004). Some of these pollutants are reaching the soil-sites (under studies) from the atmospheric pollution also.

Gupta and Ray (2004) have reported reduction in chlorophyll content, when the roots of Withania somnifera were subjected to high concentrations of Cd, Zn, Cu and Chromium (Cr) in laboratory experiments. Mishra, Tiwari and Sinha (2004) have carried out experimental studies on effects of environmental stress caused by non-material agents like energy. They have shown that even high temperature can lead to significant changes in the total haemocyte count in red cotton bug (Dysdercus koenigii).

Table 25.3 presents the important sources of water quality impairment in United States of America and Table 25.4 shows the presence of DDT in body fat of human beings in different countries. Table 25.5 lists the pesticide-concentrations observed in samples of various food materials. Though the percentage metabolic conversion product (DDE) of DDT in body of human beings in Indian population has been found to be relatively lower (37%), the total concentration of DDE is quite high due to the highest concentration (31 ppm) of DDT in body tissues. It is due to the unfortunate fact that Indian population is exposed to very high concentration of DDT in environment.

Though the presence of cumulative toxins in vegetables and fruits is also significant, relatively lower fractions of the compounds are absorbed in human body. On the other end, residues (after metabolic action in animals) present in animal products are absorbed with more relative ease in human body and consequently, the products from animals fed with or reared on polluted fodder and water may be much harmful in comparison to most of the products from vegetables, plants and trees grown even on the polluted soils and with the same water as used by the animals.

Table 25.1: Percentage of Assessed River Lengths and Lake Areas Affected.

Source of pollution	*Rivers (%)*	*Lakes (%)*	*Nature of pollutant*	*Rivers (%)*	*Lakes (%)*
Agriculture	72	56	Siltation (sediment)	45	22
Municipal point sources	15	21	Nutrients	37	40
Urban runoff/storm sewers	11	24	Pathogens	27	–
Resource extraction	11	–	Pesticides	26	
Industrial point sources	7	–	Organic enrichment DO	24	24
Silviculture	7	–	Metals	19	47
Hydrologic/habitat modification	7	23	Priority organic- chemicals	–	20
On-site wastewater disposal	–	16	–	–	–
Flow modification	–	13	–	–	–

Table 25.2: States in USA Reporting Groundwater Contamination (Out of 50)

Pollutants	*No. of States*	*Pollutants*	*No. of States*
Nitrates	49	Volatile organic chemicals	48
Petroleum products	46	Metals	45
Pesticides	43	Brine/salinity	37
Synthetic organic chemicals	36	Arsenic	28
Other substances	26	Other agricultural chemicals	23
Radioactive materials	23	Fluoride	20
Other inorganic substances	15	–	–

Fishes feeding on clams may contain 2-4 ppm DDT, while its concentration has been found to be upto 75 ppm in birds, which might eat these fishes. In man, who consumes fishes, concentration of DDT has been found to be around 15 ppm. It is estimated that in U. S. A., DDT content in human body is about 11 ppm; in England, 2.2 ppm; and from 15 to 31 ppm in Indian population, which consumes fishes (Sharma and Kaur, 2001). Since DDT is stored in body fat, it may be released massively into body fluids when the fat cells are metabolized during starvation. It is thought to get dissolved in fatty membranes surrounding nerve fibres and thus, may interfere with ion-transport in and out of the nerve membrane and may consequently hinder the passage of signals (messages) from any part of the body to brain and also in reverse direction. In case of other animals, researchers have shown that extremely low quantities of pesticides in aquatic environments may affect the productivity of organisms, kill eggs and larvae of clams, molluscs, teleosts, oysters, crustacean, fishes, etc.

Table 25.3: Important Sources of Water Quality Impairment

Rank	*Rivers*	*Lakes*	*Estuaries*
1.	Agriculture	Agriculture	Municipal point sources
2.	Municipal point sources	Urban runoff/storm sewers	Urban runoff/storm sewers
3.	Urban runoff/storm	Hydrologic/habitat modification	Agriculture
4.	Resource extraction	Municipal point sources	Industrial point sources
5.	Industrial point sources	On-site wastewater disposal	Resource extraction

Table 25.4: DDT Concentration in Body Fat of Human Beings (Sharma and Kaur, 2001)

S. No.	*Country*	*Total DDT (ppm)*	*% DDE in total DDT*	*Total DDE (ppm)*
1.	Australia	1.8	56	1.008
2.	Czechoslovakia	9.6	43	4.128
3.	Canada	4.9	68	3.332
4.	Denmark	3.3	82	2.706
5.	England	3.3	68	2.244
6.	France	5.2	67	3.484
7.	Germany	2.3	57	1.311
8.	Hungary	12.4	48	5.952
9.	India	31.0	37	11.470
10.	Israel	19.2	56	10.752
11.	Italy	5.0	64	3.200

Table 25.5: Pesticide Concentrations in Samples of Various Food Products (Sharma and Kaur, 2001)

S. No.	*Item*	*Percentage of Samples with the Residue*	*Toxin Detected*	*Range of Concentration (ppm)*
1.	Butter	15	DDT	0.3 – 8.0
2.	Cereals	38	DDT	0.25 – 1.0
			Lindane	0 – 0.40
			Endrin	0 – 1.0
			B .H. C.	0 – 0.10
3.	Eggs	60	DDT	0.25 – 1.0
			Lindane	0 – 0.40
4.	Fat	50	DDT	0 – 1.0
			Dieldrin	0 – 0.05
5.	Fruits	60	DDT	0 – 0.10
			Dieldrin	0 – 0.01
6.	Meat	20	DDT	0 – 0.20
			Dieldrin	0 – 0.01
7.	Milk	80	DDT	0.25 – 0.50
8.	Rice	4	DDT	0 – 0.10
9.	70		DDT	0 – 0.20
			Lindane	0 – 2.0
			Aldrin	0.4 – 2.0
			Heptachlor	0 – 2.0
			Chlordane	0 – 2.0
			Endrin	0 – 2.0

DDT is still being used on a large scale in tropical countries to control vector-borne diseases such as malaria and sleeping sickness (trypnosomiasis), while benzene hexa chloride (BHC) is applied as cheap insecticide in agricultural production. Aldrin and dieldrin are again being used for control of locusts and termites, and also for seed-dressings in absence of effective alternatives. Alternative non-persistent pesticides like organophosphates, carbamates and pyrethroids are costly and therefore, widespread use of cheap persistent pesticides is still going on. About 5 million people are poisoned by pesticides every year, out of which approximately 40,000 die. About 500 crop pests have developed resistance to pesticides.

At least \$ 72.5 million worth of 12 banned pesticides (also known as Dirty Dozen, *e.g.*, Camphechlor, Chlordane, Chlordimeform, DBCP, DDT, Aldrin, EDB, HCB/BHC, Paraquat, Parathion, Pentachlorophenol and 2, 4, 5–T) were exported from USA in the period from 1991–1994. Worldwide environmental and human-health costs due to pesticides are estimated between \$ 100-200 billion or \$ 12.5-25 per kilogram of pesticides used. 48 occupational pesticide poisonings cases occur every minute or about 25 million per year in developing countries.

The pesticide-fogs are transported by wind even to virgin areas like geographical regions near poles. It is found that high concentrations of persistent insecticides inhibit the tissue–development of dolphins, whales, seals and polar bears. 330 deaths occurred in Turkey due to consumption of seed-grains treated with hexachloro benzene (HCB); 80, in Columbia from consuming flour contaminated with parathion; and 17, in Mexico due to eating sugar containing parathion. It is reported in India that due to exposure to DDT, BHC, parathion, malathion, diazinon and endrin, 185, 139, 1503, 690, 889 and 1210 poisoning cases took place in States of Bihar, Karnataka, Tamil Nadu, Maharashtra, West-Bengal and in Union Territory of Delhi respectively during the last few years.

A dangerous disease known as 'Handigodu' was observed in rural areas of Chikmangalur (Andhra Pradesh) and Shimoga (Karnataka) due to faulty application of pesticides in paddy fields and coffee orchards. The people who consumed fishes, frogs and crabs from polluted rivers suffered from pain in hip and knee joints, abnormal growth of bones and unusual changes in femoral head. TCDD (2, 3, 7, 8-tetrachlorobenzo 10-dioxin) disaster at Seveso (Italy) due to an accident led to sufferings of hundreds of people. The ladies pregnant at that time gave premature births to babies with congenital deformities.

In 1970, in Hopewell (Virginia in USA), an industrial accident in Kepone-manufacturing plant exposed it's about 150 employees to the pesticide-dust in air. The employees inhaled Kepone at the rate of 3 g/m^3 and they developed symptoms of tremors and shocks alongwith other physiological problems. Later on, many of them fell victim to hepatocellular carcinoma. Due to pollution of Chesapeake Bay and James river by the same pesticide coming in effluent from a company named as Life Science Products in USA, aquatic animals including fishes showed more than 1 ppm concentration of this pesticide and consequently, fishing had to be banned in such areas in December 1975, whish practically destroyed the oyster-growing industry there. In India, pesticides very similar to Kepone are manufactured and these are marketed as chlordane, heptachlor, aldrin and dieldrin. Due to these and other such agro-chemicals, Indian food items like wheat, oilseeds, vegetables, eggs and fishes show a quite high concentrations of many pesticide-residues including DDT (Sharma and Kaur, 2001).

The results of many such studies as mentioned above may be safely extrapolated to conclude that high pollution-stresses may lead to adverse physiological effects, inflicted even on domesticated animals used for giving various food products and as a consequence, the quality of these animal products will also go down significantly. It is proved by many experimental observations that with decrease in water quality due to pollution, the biodiversity of the aquatic ecosystems also go down significantly. To quote as an example, studies have reported that Damodar River in Jharkhand and West Bengal is being slowly converted into aquatic desert due to high load of pollutants like fine coal particles, fly ash and many other pollutants from thermal power plants, steel industries, fertilizer manufacturing plants etc. (Sharma, Sinha and Kedia, 2004).

EFFECTS ON HUMAN HEALTH

Estrogens are chemical hormones naturally produced by female animals. They are carried in the bloodstream to responsive tissues where they stimulate and maintain changes that make an animal female. They are also necessary in males, but in much smaller amounts. Agents (environmental pollutants) that mimic natural estrogens are referred to as environmental estrogens or xenoestrogens. An environmental agent that can mimic one or more hormones is called an endocrine disrupter. Chemicals or other environmental agents that show estrogenic properties may be either natural or synthetic. So far, more than 250 environmental pollutants have been identified as xenoestrogens and only one, tributyl tin (TBT) as environmental androgen.

Though pesticides undergo transformation in their structure in soil through the process of biodegradation, chemical and photochemical action, significant concentrations of the original chemicals or their residues continue to persist in environment for long, especially of different agro-chemicals, which are persistent in nature.

Environmental Estrogens

Pesticides

As of late 1993, 35 of the 45 chemicals identified as having estrogenic properties were pesticides. The insecticide o, p-DDT, widely used from 1943 until 1973, is the best known of these. o, p-DDT was also the first chemical implicated as having adverse effects on animal reproduction; in some bird species, including bald eagles and osprey, it thinned egg shells; it also affected reproduction in bird species whose egg shells were not affected. A major population drop seen in western sea gulls was attributed to DDT exposure. Many so-called lesbian gulls shared nests with other females and produced eggs that developed into males with feminized reproductive tracts; male gulls showed feminine characteristics and some were even sterile.

Another major example of adverse effects on animal population occurred after a spill of the pesticide dicofol into Florida's Lake Apopka in 1980. The dicofol contained about 15% DDT as a contaminant. Later, it was observed that 95 % of Lake Apopka's alligator eggs were failing to hatch, as compared to a normal hatch rate of 70-80%. Half of those that did hatch died within 2 weeks and those that survived had feminine characteristics. These effects were traced to pp- DDE, a degradation product of DDT. DDE is not a "true" xenoestrogen because, rather than acting as an oestrogen, it blocks the action of androgen, the male hormone.

Pesticides with estrogenic properties have low potency as compared to natural animal estrogens. The effects on birds and alligators noted earlier occurred in the presence of high doses. The high DDT concentration associated with eggshell thinning resulted from its build-up in the environment over the 30 years in which it was used, and the high DDT concentrations in Lake Apopka resulted from a spill. Since DDT was banned in 1973, its levels in the American environment and in animal and human blood and tissues have declined.

A 1996 report raised the possibility of synergistic effects of these pollutants. It presented data showing that any one of four chlorinated pesticides that were investigated had only weak estrogenic properties. However, when two were tested together, the effect was synergistic. This finding enhanced concerns that xenoestrogen could have effects even at low environmental levels. In addition, one of the four pesticides tested, endosulfan, is still used in the United States of America (Vonier, 1996; Rollanad, 1995; Guillette, 1994; Smith, 1996).

Polychlorinated Biphenyls (PCBs)

Polychlorinated biphenyls (PCBs) are another group of chemicals implicated in reproductive problems. By the time they were banned in the 1970s, 2 to 3 billion pounds of PCBs had been produced and used. Many sites around the Great Lakes were among those highly contaminated with PCBs. Fish in or near these sites were also highly contaminated. In one study, when the fish constituted 20% of laboratory-raised mink's diet, the mink had a reduced number of kids with a reduced survival rate; when the fish were 40% of their diet, the mink did not reproduce at all. In the twenty years, since PCBs were banned, their concentrations in Great Lakes fish have dropped as much as 90%, although sensitive animal species such as mink still suffer.

Another example of adverse effects attributed to PCBs occurred in terns nesting off the Massachusetts coast. These birds attracted attention because of the abnormal behavior of nesting females. As had occurred with gull populations exposed to DDT, female terns were found sharing the same nest. Investigations revealed that some male embryos in the eggs of these terns had both male and female tissue, or improperly developed male organs. Their nesting area was near a site highly contaminated with PCBs.

The greatest route of human exposure to PCBs is fish consumption. Concern arose as to whether PCB levels in these fish could affect human reproduction. In a study designed to examine this concern, children born to women who ate 2 to 3 Lake Michigan fish a month were studied. The babies of these women were reported to have lower birth weight and, at age 4, to exhibit short-term memory problems. Human populations have clearly shown adverse effects in response to higher doses.

In the late 1970s, in Japan and Taiwan, thousands of individuals were poisoned after ingesting rice oil accidentally contaminated with PCBs. Women who were pregnant at the time had an abnormal number of foetal deaths and stillbirths. Babies born to women affected by the Taiwanese incident had lower birth weight, abnormal skin pigmentation and erupted teeth. They later showed lower IQ scores than a control group of children and more behavioral problems. Although environmental levels have greatly decreased, PCB levels in human blood and tissues have declined more slowly than those of DDT. In fact, environmental levels of PCBs appear to have reached a plateau and are now declining much more slowly than previously.

Although the process is expensive, PCBs in land-based hazardous waste sites can be cleaned up. It is more difficult and expensive to clean up the contaminated sediments where PCBs, because of their low water solubility, got concentrated. In sediments, they are sheltered from the degrading effects of sunlight and oxygen (Vonier, 1996; Rollanad, 1995; www.mel.tulane.edu/ecme/eehome.basic; Stephen, 1995; www.C3.org/chlorine_iss ue/health/nov.html).

Dioxins

Dioxins are a family of chlorinated chemicals related to PCBs. They have powerful biological effects and can interfere with the functioning of a number of hormones. Animals are particularly sensitive to dioxin exposure when it occurs in the uterus or shortly after birth. In one-well known study, pregnant rats that were given one tiny dose of the most toxic dioxin, 2, 3, 7, 8-TCDD, produced male offspring that demonstrated reduced sperm production and other adverse effects. However, unlike PCBs and DDT, dioxins have antiestrogenic properties in some situations, and some less toxic dioxins are being evaluated as potential treatments, for oestrogen-sensitive breast cancers. Considering their extreme toxicity, it is fortunate that environmental levels of dioxins are far lower than those of other polychlorinated chemicals. This is because, accepting tiny amounts synthesized in research laboratories, dioxins are not deliberately produced.

Dioxins form as contaminants during combustion, especially combustion of MSW and medical waste. Forest and grass fires also generate dioxins, as do other combustion sources. Once emitted to the atmosphere, the dioxin particulates drift onto water and land. They settle onto vegetation that is eaten by cattle and other animals and concentrate in the animals' fat and milk. Dioxins also form as contaminants during the chlorine bleaching of wood pulp used to make paper. Released in bleaching effluents into rivers, dioxins concentrate in sediments, from which they are taken up by invertebrate animals that are subsequently eaten by fish. Birds and mammals, including humans, are exposed when they eat contaminated fish.

Except for people who eat larger than average amounts, fish account for about 7% of total dioxin exposure. Most of the remaining exposure results from consumption of meat and dairy products. A vegetarian diet can lower dioxin intake by about 98%. In the homes of smokers, burning cigarettes emit dioxins, tiny in amount, but enough to double the smoker's dioxin exposure. Dioxins are also generated by home wood burning and accumulate in wood soot. MSW and medical waste combustors are the major known sources of dioxin emissions, although their emissions vary greatly. Dioxin amounts released by medical waste combustors varies from 117 to 450 ng/m^3 and that released by municipal solid waste (MSW) combustors from 1 to 10,700 ng/m^3.

Dioxins were also previously formed as contaminants during the manufacture of chlorinated pesticides, but modified processes have greatly reduced their formation. The process used to bleach wood pulp with chlorine has also been modified since the mid-1980s, resulting in reduction of dioxin discharges to water upto 95%. Even at the tremendously sensitive detection level of 10 parts per quadrillion, 2,3,7,8-TCDD is not ordinarily detected in bleaching effluents. In mills that switch from elemental chlorine to chlorine dioxide as a bleaching chemical, effluents do not contain detectable dioxin. Some call for elimination of any chlorine-containing chemical in pulp bleaching with the goal of removing any remaining dioxin emissions. Presumably as a result of increased controls placed on dioxin-emitting sources, dioxin contamination of the environment has decreased by about half in the past 20 years (Davies and Bradlow, 1995).

Polycyclic Aromatic Hydrocarbons (PAHs)

The PAH family of chemicals is related to PCBs and dioxins, but does not contain chlorine. Some are known human carcinogens and others show estrogenic properties. PAHs are also environmentally persistent substances that bioaccumulate in animal fat. Like dioxins, almost any combustion process including internal combustion in automobiles produces PAHs. However, whereas dioxins are primarily produced during waste combustion, the major source of PAHs is fossil fuel combustion, although some are formed in wood-burning fireplaces and stoves. Charcoal-broiled foods especially, but also oven-browned foods and toast, contain these chemicals. PAHs cannot be eliminated from the environment, but they can be controlled. PAHs emissions from coal burning are down as compared to historical releases and the same is true for the motor vehicle emissions.

PAHs produced in home can be reduced by preventing grease from dripping onto charcoal-broiled foods and preventing fried or baked foods from becoming too brown. Fireplaces and stoves need regular maintenance to minimize soot formation and emissions. Smoking cigarettes within the home need be eliminated. However, unlike PCBs, polychlorinated pesticides and dioxins, which society either has eliminated or strictly controlled, PAHs emissions, although reduced, continue as a result of our ongoing dependence on fossil fuels (Davis and Bradlow, 1995).

XENOESTROGENIC CHARACTERISTICS OF SOME UBIQUITOUS CHEMICALS

A number of chemicals associated with plastics and plastic uses have estrogenic properties and are pervasive in the environment. Cans coated inside with a plastic resin leak the weakly estrogenic chemical bisphenol-A into the food. Phthalates, chemicals used to make plastics flexible, leach from plastics. Plastics also shed Nonylphenols. The chemical DBP is also found in plastic plumbing pipes. Weakly estrogenic preservative BHA is an antioxidant used as a food additive. The metals mercury, cadmium and lead show estrogenic activity (Lidhutts, 1996; Smith, 1997).

SANITARY SEWAGE

Sewage effluents contain chemicals with estrogenic activity. Many fish in English river, caught in the lagoons and pools just below the discharge from sewage plant looked quite bizarre. Even experienced fishermen could not identify whether a fish was male or female. The fish showed male and female characteristics at the same time. Researchers reported testing 20 chemicals in sewage effluent, chosen at random from amongst the thousands of chemicals present. Half of these were detected to be weakly estrogenic (Lidhutts, 1996).

Due to such pollutants and their high concentrations, many diseases have gone to a very high prevalence rate. In addition to breast cancer, endometriosis is another disease whose incidence has risen in women. Again, xenoestrogens are suspected to be the possible cause. In endometriosis, cells that normally line the inside of the uterus are found in other locations in the body. This causes painful symptoms and can lead to infertility (Davis and Bradlow, 1995).

The ability of a hormone to act at extremely low doses makes the finding of environmental estrogens disturbing. However, as with most other chemicals, a hormone has a threshold. At yet higher concentrations in a living animal, a negative feedback comes into play and turns off the system. From this perspective, the truism, "The dose makes the poison," is true only up to a point for hormones. One major problem in evaluating the possible effects of endocrine disrupters including xenoestrogens is the difficulty of meaningfully evaluating chemicals for hormone-mimicking behaviour.

Evaluating potential endocrine disrupters is even more difficult. Nonetheless, future evaluations of chemical's toxicity will need to include an assessment of its ability to act as an endocrine disrupter also. The information that has emerged on xenoestrogens and other chemicals with hormonal activity is complex and may become more so before answers are sorted out of the confusion (Arnold, 1996; Cody, 1995; Stenbbraber, 1996).

SOME REMEDIAL MEASURES

Certain important measures to be adopted for reduction of risks due to pesticide-pollution at present are as listed below:

(i) First of all, indiscriminate and excessive use of pesticides needs to be stopped by spreading awareness about the risks involved and other aspects related to safe application of the pesticides.

(ii) An effective Continuing Monitoring and Information System for presence of various pesticides in environment must be established.

(iii) The interactions phenomena of biocides with other chemicals like detergents, and domestic and industrial wastes need be established to understand any 'synergistic' and 'antagonistic' behaviours which may be existing amongst them.

(iv) Less persistent and biodegradable pesticides should be used as far as possible.

(v) Biological pest control and herbal pesticides should be encouraged to the maximum possible extent.

(vi) Solid-wastes and effluents (liquid and gaseous) from pesticides-manufacturing industries must be treated properly before their release into the environment.

(vii) Biocides should be applied only when the population(s) of harmful organism(s) exceed(s) the economic threshold limit(s).

(viii) Domesticated animals should not be allowed to graze on the treated crops/grasses.

(ix) Appropriate Central or State Government Agency should regulate the formulation, production, packaging, handling, distribution and the aspects related to application of various biocides.

(x) Agricultural produces should be harvested after the waiting period (after application of biocides) becomes over.

(xi) Direct mixing of biocides with food items must be avoided under all circumstances.

(xii) Consumers should wash thoroughly with water and peel all the vegetables and fruits before use.

CONCLUDING REMARKS

In fact, at present, all the three components of biosphere have detectable presence of agrochemicals, particularly various types of biocides. The air, used for breathing, water used for drinking and soil used for growing vegetables, plants and trees are polluted with these chemicals and consequently, our food items are showing the presence of pesticides to objectionable limits. The pesticides though principally applied on plants migrate to other components of biosphere. They enter aquatic eco-system through rainfall (picked up from air and spray drift), runoff from agricultural fields and sanitary (domestic and industrial) sewage fluxes, etc.

Time has now come to shift from short-term techniques for higher production to carefully balanced long-term goals of sustainable agricultural practices alongwith environmental protection. To achieve this goal, biological control of pests should be encouraged and genetic engineering or biotechnology should be employed to impart resistance to economic plant kingdom to fight against various diseases and insects.

In addition to all the adverse consequences of pollution on quality of animal products discussed in earlier sections, there is another impact in terms of reduction in capability of Indian dairy-products to compete with such products from other countries in International Market due to higher concentrations of various pollutants in such products made in India. This has emerged as a terrific challenge to marketability of Indian dairy-products in present era of globalization and liberalized economies.

REFERENCES

Arnold, F. (1996), Synergistic Activation of Estrogen Receptor with Combinations of Environment Chemicals, Science Vol. 272, pp. 1489-1492.

Arya, S. K., Singh, P. and Ray, B. K. (2004), Responses of Vicia faba L. to Copper Stress, Abstract Proceedings of National Seminar on Strategies for Environmental Quality Management in Urban Areas, Sri AKAPG College, Varanasi, October 28, pp. 89-90.

Cody, P. (1995), Environmental Estrogens: Global Problem, Detailed Report Available at the website: *www.desaction.org*

Davis, D.L. and Bradlow, H.L. (1995), Can Environmental Estrogens Cause Breast Cancer? Scientific American, Vol. 273, No. 4, pp. 166-172, Detailed Report Available at the website: *du/ecme/eehe*

Guillette, L. (1994), Development of Abnormalities or the Gonad and Abnormal Sex Hormone Concentrations in Juvenile Alligators from Contaminated and Control Lakes in Florida, Environmental Health Perspectives, Vol. 102, No. 8, pp. 680-688.

Gupta, A. K. and Ray, B. (2004), Bioaccumulation of Cadmium, Zinc, Copper and Chromium by Withania somnifera, Abstract Proceedings of National Seminar on Strategies for Environmental Quality Management in Urban Areas, Sri AKAPG College, Varanasi, October 28, p. 21.

Gupta, P. (2004), Estimation of Heavy Metal Toxicity in Plants Grown on Polluted Sites, Abstract Proceedings of National Seminar on Strategies for Environmental Quality Management in Urban Areas, Sri AKAPG College, Varanasi, October 28, pp.18-19.

Jefry, P. K. (1999), 'Effects of Cigarette Smoke and Air Pollutants", in Air Pollution and Health, Holgate, S. T., Samet, J.M., Koren, S.H. and Maynard, R.L. (Eds.), Academic Press, New York, pp. 219-268.

Lata, S. and Srivastava, V. M. S. (2004), Histopathological Changes in Fresh Water Fish Under Stress of Some Chemicals Used in Carpet Dyeing Industry in U. P., Abstract Proceedings of National Seminar on Strategies for Environmental Quality Management in Urban Areas, Sri AKAPG College, Varanasi, October 28, pp. 15-16.

Lidhutts, L. (1996), The Debate Continues Over the Possible Decline in Sperm Quality, Fertility and Sterility, European Journal of Cancer, Vol. 65, No. 5, pp. 909-911.

Mishra, A., Tiwari, R. K. and Sinha, B. K. (2004), Effect of High Temperature on Haemocyte Population in Red Cotton Bug (Dysdercus koenigii), Abstract Proceedings of National Seminar on Strategies for Environmental Quality Management in Urban Areas, Sri AKAPG College, Varanasi, October 28, p. 26.

Pandey, A. and Parveen, E. (2004), Water Pollution through Agricultural Discharge, Abstract Proceedings of National Seminar on Strategies for Environmental Quality Management in Urban Areas, Sri AKAPG College, Varanasi, October 28, p. 41.

Prakash, A. and Srivastava, V. C. (2004), Impact of Urban Sewage Sludge on Leguminous Plants, Abstract Proceedings of National Seminar on Strategies for Environmental Quality Management in Urban Areas, Sri AKAPG College, Varanasi, October 28, pp. 17-18.

Rollanad, M. (1995), Environmentally Induced Alterations in Development Focus on Wild Life, Special Issue of Environmental Health Perspectives, Vol. 103, pp. 101-109.

Sharma, B. K. and Kaur, H. (2001), Environmental Chemistry, Third Edition, GOEL Publishing House, Meerut, pp. 52-59 and 98-121.

Singh, D. (2004), Studies on Utilization of Treated Domestic Sewage for Irrigation, Abstract Proceedings of National Seminar on Strategies for Environmental Quality Management in Urban Areas, Sri AKAPG College, Varanasi, October 28,, pp. 49-50.

Smith, E. (1997), Phytoestrogen, Some Good and Some Bad Phytoestrogen, Naturally Occurring Plant Estrogens, The Ones to Avoid, detailed report available at the website: *www. fibrocytic.com*

Smith, L. (1996), How Exposure to Synthetic Environmental Estrogen May Lead to Decline in Wild Life Populations, Detailed Report Available at the website: *www.msec.uncc.edu/Environmental %20Estrogen.doc*

Srivastava, A. and Singh, H. N. (2004), Effects of Physical and Chemical Mutagens on Vigna imbellate, Abstract Proceedings of National Seminar on Strategies for Environmental Quality Management in Urban Areas, Sri AKAPG College, Varanasi, October 28, p. 40.

Stenbbraber, S. (1996), Xenoestrogen and Woman's Health, Detailed Report Available at the website: *www.Infertillityworkshop.com/articles/hormonalhealthxenoestrogen.htm*

Stephen, H. (1995), Environmental and Dietary Estrogens and Human Health: Is There a Problem? Environmental Health Perspectives, Vol. 103, No. 4, pp. 346-351.

Vonier, P. (1996), Interaction of Environmental Chemicals with the Estrogen and Progestrogen from the Oviduct of American Alligator, Environmental Health Perspectives, Vol. 104, pp. 104-112.

www.C3.org/chlorine_issue/health nov.html: Detailed Report Available About Environmental Endocrine Issue.

www.mel.tulane: Detailed Report Available About Environmental Estrogens and Other Hormones.

Yadav, A. and Shukla, Y. (2004), Carcinogenic and Co-carcinogenic Potential of Cypermethrin on Mouse Skin, Abstract Proceedings of National Seminar on Strategies for Environmental Quality Management in Urban Areas, Sri AKAPG College, Varanasi, October 28, p. 38.

Bibliography

1. Aftab Alam (ed.), *Human Rights in India-Issues and Challenges* Raj Publications, New Delhi.
2. Anjaneyulu .Y, **'Environmental Impact Assessment Methodologies'**, Ed. by CA Sastry, 35 Publications, Hyderabad, 2002.
3. Barry J., **'Rethinking Green Politics'**, Sage Publications, New Delhi, 1999.
4. Chadha, SK (ed.), **'Environmental Crises in India'**, International B.K Distributers, Dehradun, 1992.
5. Cunningham,William P and Cunningham Marry Ann,(ed.) **'Principles of Environmental Science, Inquiry and Applications'**, Tata McGrew Hill Publishing Corporation, New Delhi, 2004.
6. Dash, M.C, **'Fundamentals of Ecology'**, Tata McGrew Hill , New Delhi, 1993.
7. Deepender Basu (ed.), **'Environment and Ecology: The Global Challenge'**, Printwell Publications, Jaipur, 1995.
8. Digvijay Singh, **'The Eco-Vote People's Representatives and Global Environment'**, Prentic Hall of India Private Limited, New Delhi-1985.
9. Divan, AP and Arora, DK., **'Elements of Ecology'**, Anmol Publictions, New Delhi, 1995.

10. Gadgil, Madhav and Ramachandra Guha, **'Ecology and Equity; The Use and Abuse of Nature in Contemporary India'**, Routledge, London, 1995.

11. Gaston, Kevin J and Spicer, John I, **'Bio-Diversity: An introduction'**, Oxford Blackweell Publications, 2004.

12. Khan, TT and Shishodia, YS (ed.), **'Bio-diversity, Conservation and Sustainable Development'**, Pointer Publishers, Jaipur, 2005.

13. Leveque, Chiristian and Mounolou, Jean-Claude, **'Bio-Diversity'**, Chicherter, John Wiley, 2003.

14. Nivedita Menon (ed.), **'Gender and Politics in India'**, Oxford University Press, London, 1999.

15. Pahari Basu (ed.), **'Environment and Society'**, Reliance Publishing House, New Delhi, 1997.

16. Peter Borrelli (ed.), **'Cross Roads Environmental Priorities for the Future'**, Island Press, Washington D.C, Covelo CA, 1988.

17. Pravin Sheth, **'Environmentalism Politics, Ecology and Development'**,Rawat Publications, Jaipur and New Delhi ,1997.

18. R. Guha, **'Environmentalism: A Global History'**, Oxford Publications, New Delhi, 1999.

19. R.N Basu, **'Compendium of Terms in Ecology and Environment'**, Naya Udyog, Kolkata, 2004.

20. S. Krishna, **'Environmental Politics'**, Sage Publications, New Delhi, 1996.

21. S. M. Khopkar, **"Environmental Pollution Analysis,"** New Age International (P) Limited, New Delhi-2001.

22. Salpeker, Aradhana and Kadambari Sharma, **'Encyclopedia of Ecology and Environment'**, Bharti Khadi Gramodyog, Vikas Abhikaran, New Delhi, 2006.

23. Samar Singh, **"Religion and Conservation"**, World Wide Fund for Nature-India, New Delhi-1999.

24. SK Agarwal, 'Environmental Scenario for 21st Century', APH Publishing Corporation, New Delhi, 2003.

25. T O'Riordan, *Environmentalism*, Pion Limited, 207 Brondesbury Park,London.

26. T.P Sankaran Kutty Nair (ed.), **'Society and Politics-Issues and Perspectives'**, Indian Institute of Police Studies, Thiruvananthapuran, Kerala, 1998.

27. Vandana Shiva (ed.), **'Minding Our Lives, Women from the South and North reconnect Ecology and Health'**, Kali for Women , New Delhi,1993.

28. Vandana Shiva in association with J.Bandyopadhyay, Pandurang Hegde, B.V.Krishnamurthy, John Kurien, G.Narendranath, Vanaje Ramprasad, S.T.S Reddy, **Ecology and the Politics of Survival: Conflicts over Natural Resources in India**, Sage Publications,New Delhi,1991.

29. Vandana Shiva, **'Staying Alive Women, Ecology and Survival in India'**, Kali for Women, New Delhi, 1988.

30. V S Mahajan (ed.), **'Environmental Protection: Challenges and Issues; Impact of Rapid Development and Policy Measures'**, Deep & Deep Publications, New Delhi, 1993.

Website

1. www.environment.about.com
2. www.globalwarming.com
3. www.worldclimatereport.com
4. www.mkgandhi.org